Genomics and Democracy

Life sciences, ethics and democracy 1

Series editors:

Genomics and Democracy

Towards a 'Lingua Democratica' for the Public Debate on Genomics

Edited by
Peter Derkx and Harry Kunneman

Rodopi Amsterdam-New York, NY 2013

Cover photo by Alexander Hardwick

Photo editing Kate van Loon

Cover design: Studio Pollmann

The paper on which this book is printed meets the requirements of "ISO 9706:1994, Information and documentation - Paper for documents - Requirements for permanence".

ISBN: 978-90-420-3719-9
E-Book ISBN: 978-94-012-0975-5

Printed in the Netherlands

Contents

Part 3: Corporations, Governments and Democracy

Introduction

*Harry Kunneman & Peter Derkx**

This book explores the contexts and consequences of scientific and technological progress in the life sciences, in view of possibilities to further democratic debate with regard to this progress. To this end it focuses on the idea of a *lingua democratica* for the life sciences, in particular the field of genomics. Developments in this field clearly illustrate the new political, social and ethical challenges flowing from the spectacular progress of the life sciences during the last decades. The new technological possibilities and social practices connected with these developments intrude into domains that for a long time have been the provenance of religion and ethics. There they touch upon deep-seated convictions and emotions. Moreover they are strongly influenced by economic and political interests. Thus they can no longer be adequately conceptualized along the lines of a more or less autonomous scientific development, based on a disinterested search for true knowledge, providing an objective knowledge base which interacts with values and interests only on the level of practical application. Especially in the life sciences, questions of scientific truth and scientific progress are getting more and more *mixed up* with questions regarding values and interests, as testified clearly by the heated debates on the possibilities, benefits and dangers of genomics research.

The contributions to this book start from the premise that in this situation neither clinging to the idea of value-neutral, disinterested science, nor the complete abandonment of this idea in favour of postmodern relativism will be of much help. Instead the different contributions to this book explore the idea of a 'lingua democratica' for the life sciences and sketch the contours of this notion by focusing on the field of genomics. In the development of the life sciences during the last decades, many different stakeholders are involved: not only scientists, technologists and clinicians, but also corporations, governments, NGO's, representatives of secular and religious worldviews and last but not least ordinary citizens. In view of this plurality of different stakeholders the idea of a lingua democratica refers to the development of a shared language, allowing for open and deliberative forms of argumentation between these different stakeholders. The history of this notion provides a vivid illustration of the complex interactions between scientific, technological, political and ethical questions which the idea of a 'lingua democratica' tries to address.

The Netherlands Genomics Initiative

The idea of a 'lingua democratica' has come to the fore in the context of the efforts of Dutch government agencies to stimulate genomics research in the Netherlands. In 2001 the Dutch Government decided to subsidize scientific research in the field of genomics with an amount of half a billion Euro. This decision was not only motivated by the fear that scientific research in this field would lag behind other European nations and especially the United States, but also by the expectation of great future benefits. According to a government committee on the development of a Genomics Knowledge Infrastructure, research in the field of genomics was "of strategic importance for the quality of society in the Netherlands. Health care can be improved, due to more effective combating and prevention of disease. Food can be made healthier and safer, with a longer shelf life. And genomics will allow for development of more sustainable methods of production."[1]Apart from these social benefits, genomics research also had a clear economic importance. As explicated in more detail in chapter 10 by Alle Bruggink, following the decision of the European Union that government subsidized research collaboration between universities and industry should be pre-competitive, a close cooperation had developed in the eighties between big Dutch corporations and universities, This decision allowed for the development of new forms of cooperation without blurring too much the differences between academic interests and commercial applications, at least in the perception of the principal stakeholders involved. To further optimize the cooperation between the academic, industrial and social parties involved, a Nationaal Regie Orgaan

[1] NGI, 2007, p. 5

Genomics (Netherlands Genomics Initiative) was established by the government in 2002. This NGI defined genomics as:

> ..the mapping of the genes of humans, animals, plants and micro-organisms, by means of DNA sequencing, and wide-scale research into the functions of genes and the manner in which hereditary characteristics contained in cells are translated into the function of a cell and ultimately into the function of the entire organism. This also includes high-throughput technologies, such as proteomics and metabolomics, and the bioinformatics that enable the data processing and analysis of colossal amounts of data. Genomics is a set of technologies that has become an indispensable tool in the current field of life science research.[2]

Right from the beginning the social aspects of genomics research were a matter of clear concern for the NGI, although it was also clear that this concern was not completely disconnected from the interests of the research consortia that had won the competition for the big amount of research money involved. It was evident from the start that these research consortia, in which universities and different industrial partners collaborated closely, were completely convinced of the great social benefits their research would bring, while the general public in the Netherlands was much more sceptical about the benefits genomics research could bring, and in part even deeply worried about the risks connected with 'genetic engineering'. This mixed concern resulted in a call for research proposals on 'Social aspects of genomics research', subsidized with 5 million Euro by the Netherlands Organization or Scientific Research, followed a few years later by the establishment of a Centre focusing on the interaction between 'genomics' and 'society'. On its website the NGI designates the activities of this Centre for Society and Genomics as follows:

By stimulating (genomics) researchers to assess their work from a broader societal perspective and by providing them with new insights and skills, the NGI Centre for Society and Genomics (CSG) aims to improve the interaction between 'genomics' and 'society' and enhance the quality of the debate. NGI's strategy within the 'Society' theme is

---

[2] NGI, 2007, p. 4

twofold. One is stimulating academic research involving social sciences, law, ethics, philosophy and economy through the funding of targeted research projects within a clear domain: food, health or environment. The other involves developing a broad array of public communication and education activities. These activities range from practical introductions to genomics in high schools and participation in science festivals, to online discussions with specific target groups (parents, patients, farmers, etc.) and public debates organised around current topics.[3]

A *lingua democratica* for the genomics debate

The notion of a 'lingua democratica' for genomics was developed as a critical contribution to the NWO research program on *Social Aspects of Genomics*. In this program a heavy emphasis was laid on the furtherance of reflexivity, responsibility and broad participation with regard to the development of genomics. In the call for proposals, this intention was further specified in terms of the desirability of a *lingua franca* which could further interactions between the wide range of different scientific disciplines involved in genomics research and could moreover help to bridge the gap between the specialist languages of the scientists involved and the concerns of the wider public with regard to the possibilities and the risks of genomics. This was an interesting idea, but in our view it was also potentially misleading. In our proposal for a research project on the *Social Aspects of Genomics*, we wrote:

> The notion of a *lingua franca* presupposes the possibility of a shared language between the different parties involved - such as scientists, corporations, representatives of specific world views, NGO's and the general public - which would allow them to transcend the different world views and presuppositions embodied in their respective local languages. However, these 'local languages' are not only characterized by different presuppositions with regard to the many complex questions at stake in the

[3] Since November 2011 the centre is called CSG, Centre for Society and the Life Sciences

> debates on genomics, but also express divergent but legitimate *interests.* Instead of trying to transcend these divergent interests by means of a hypothetical *lingua franca*, this research project aims to elaborate and concretize the idea of a *lingua democratica* for genomics. In such a *lingua democratica* the differences between the central presuppositions and core concepts of the parties involved are seen as a legitimate expression of the complexity of the questions at stake. Given this complexity, the communication and interaction between the stakeholders involved can be furthered by respecting the differences but looking at the same time for conceptual 'interfaces' and points of connection between the central concerns and presuppositions of the different languages. These languages themselves do not have a unitary character but exhibit different articulations, some more closed and 'self-centered', others more open towards interaction and possible connections with the concerns and presuppositions of other languages involved in the debates on genomics. This research project aims to elucidate the notion of a *lingua democratica* for genomics and contribute to its development by identifying and elaborating 'interaction-prone' articulations of the language of central parties to the debates on genomics. More in particular the research project focuses on specific articulations of these languages offering opportunities for *public* deliberation and democratic debate with regard to the many pressing questions connected with the further development of genomics.

Our proposal comprised five projects, focusing on the identification and further development of more or less 'open' or 'interaction-prone' versions of the perspectives of principal parties to the debates. Because the researchers in our team came from different disciplinary backgrounds - biology, engineering, philosophy, sociology and ethics – we faced the problem of developing a lingua democratica also on the micro-level of our research team. In the course of our discussions three broad conceptual orientations came to the fore that not only helped us to provisionally bridge the disciplinary differences within our team, but also served as pointers for the conceptual development of the idea of lingua democratica for the life sciences:

a) The first pointer was provided by recent developments in the field of complexity theory, especially the emergence of critical complexity thinking, as exemplified among others in the work of Paul Cilliers, Edgar Morin, Mario Giampietro and Robert Ulanowicz.
b) The second broad orientation helping us to further develop the idea of a lingua democratica, was provided by social constructionism, by the distinction between Mode 1 and Mode 2 knowledge production introduced by Gibbons et al., and by the related idea of 'post-normal science', introduced by Ravetz and Funtowicz.
c) In the third place the shared frame of reference which gradually emerged during our research was oriented by new perspectives in the domain of social and political theory, highlighting the far reaching changes which have taken place during the last decades in the internal structure of democratic, 'post-industrial' societies and the global networks they partake in. Especially helpful for us in this respect were deliberative models of democracy, as developed among others by Sheila Benhabib, Jürgen Habermas and Axel Honneth.

Ad a) Critical complexity thinking

A first, formidable obstacle for the development of a lingua democratica for the life sciences is of course the deeply entrenched idea that scientific knowledge should be objective and value free. This idea implies that values and interests only come into play when scientific knowledge is applied in practical contexts. When this idea would hold water, the whole notion of a lingua democratica for the life sciences would be void, because in that case one of the languages involved would be above critique and not in need of any form of enriching interaction with other parties. However, the complete abandonment of the idea of objective knowledge, as advocated by some postmodern thinkers and by radical constructivists, would also block the development of a lingua democratica. When the idea of objective knowledge is reduced to an ideology masking the power relations supposedly underlying all claims to truth and objectivity, then all scientists and technologists who are deeply committed to the value of objectivity are ruled out beforehand as serious participants to the debate. This would contradict one of the central premises underlying the notion of a lingua democratica, namely the idea that the differences between the central presuppositions and core concepts of the different parties involved should be considered as legitimate expressions of the *complexity* of the questions at stake. Respecting these differences is thus an indispensible starting point for the development of a lingua democratica. Respect however is not the same as blind acceptance: it allows and even asks for critical questioning and suggestions. An important point of entrance for such critical questions and suggestions is provided by the fact that the languages of the different parties involved do not have a unitary character, but exhibit different articulations, some more closed and 'self-centered', others more open towards interaction and to possible connections with the concerns and presuppositions of other parties. At this point the development of a lingua democratica for genomics (and for the life sciences in general) can profit from the gradual emergence of a new perspective on science and objectivity during the last decades, which travels under many names but can perhaps best be designated - with Paul Cilliers (1998) - as critical complexity thinking. A good entrance to the change of perspective brought about by critical complexity thinking in comparison with the mainstream view on science and objectivity, is provided by a recent book by ecologist Robert Ulanowicz: *A Third Window: Natural Life beyond Newton and Darwin*.[4]

In this book Ulanowicz tries to integrate recent developments within systems ecology, relational biology, complexity theory and the philosophy of science. His central claim concerns the opening up of a third 'window' on nature within contemporary life sciences, complementing the first, 'Newtonian' window, and the second 'Darwinian' window. Looking out from the first two windows one perceives a deep opposition between on the one hand the 'hard' natural sciences, focusing on inanimate matter and oriented to causal explanations based on objective empirical research, and on the other hand the 'soft' humanities oriented towards value ridden interpretative processes[5]. However, seen from the third window important continuities between them come into view. Contemporary life sciences do not only deal with causal laws but also with 'complex singularities', as Ulanowicz has it, that is to say: with unique forms of interference and co-creation between complex constellations which cannot be predicted with certainty beforehand, but have an emergent

[4] Ulanowicz, 2009

[5] Broad, 1930

character, such as the advent of a new species or the emergence of language among the first human primates.

According to Ulanowicz, the presence (or rather predominance) of complex singularities in nature and the probability that such singularities will occur again, imply the acceptance of the fundamental openness of our future. This is an important conclusion in view of the development of a lingua democratica for the life sciences, because it applies not only to the life sciences themselves, but also to dogmatic metaphysical, political and religious perspectives claiming an absolute basis for their own perspective. Looking out from Ulanowicz third window we see not only the restrictions of Newtonian forms of determinism and of Neo-Darwinian, gene-oriented versions of determinism, but also the inadequacy of all religious, philosophical and political worldviews promising some form of metaphysical certainty. As Ulanowicz says:

> …all who look out the third window perceive in the distant background an 'epistemological veil' that neither rationality nor empiricism can penetrate. Whether that veil shrouds emptiness or obscures a higher agency remains perforce unknowable. Radical uncertainty is the portion of *all* who inhabit this mortal coil. In such a world, patience with one another becomes a clear virtue[6].

In this respect the perspective developed by Ulanowicz shows a clear resemblance with the view developed by Paul Cilliers in his book *Complexity and Postmodernism.*[7] According to Cilliers, the Newtonian worldview is applicable only to simple systems such as our planetary system. When we are confronted with complex systems however, such as living entities and their intricate interactions within ecosystems, we have to recognize that our models of such complex systems necessarily involve simplifications:

> The knowledge we have of complex systems is based on the models we make of these systems, but in order to function as models [..] they have to *reduce* the complexity of the system. This means that some aspects of the system are always left out of consideration. The problem is

[6] Ulanowicz, 2009, p. 163

[7] Cilliers, 1998

> compounded by the fact that that which is left out, interacts with the rest of the system in a non-linear way and we can therefore not predict what the effects of our reduction of the complexity will be, especially not as the system and its environment develops and transforms in time.[8]

According to Cilliers this line of reasoning justifies the conclusion that modeling complex systems necessarily implies choices which cannot be justified on the basis of 'objective facts' alone. Instead they also bring normative and ethical considerations into play:

> The failure to acknowledge the complexity of a certain situation is not merely a *technical* error, it is also an *ethical* one.... We cannot make purely objective and final claims about our complex world. We have to make choices and thus we cannot escape the normative or ethical domain... Normative issues are intertwined with our very understanding of complexity.[9]

According to Mario Giampietro this intertwinement implies the necessity to *combine* descriptive and normative forms of 'quality control' when modelling complex systems.[10]

At this point, the differences between the critical perspective on complexity developed by Ulanowicz, Cilliers, and Giampietro on the one hand and complexity *theory* as developed for instance within the Santa Fe school, become clearly visible. Whereas complexity theorists such as Stuart Kauffmann and John Holland try to uncover the general laws underlying the functioning of all complex systems,[11] Cilliers clearly perceives the inherent limitations of all models of complex systems, including those developed by complexity theorists. Thus Cilliers, Ulanowicz and Giampietro underline the normative and ethical questions coming into play when we try to model the complexity of for instance ecological systems, of language or of social relations. As Cilliers explicates in more detail in chapter 1 of this book: 'normative issues are intertwined with our very understanding of complexity.'

---

[8] Cilliers, 2005, p. 256
[9] Cilliers, 2005, p. 259/264
[10] Giampietro et al., 2012
[11] Kaufmann, 1993, 2000; Holland, 1998

Ad b) Social constructionism and the emergence of new forms of knowledge production

Social constructionism provided us with a second fruitful perspective to address these normative issues with the help of the idea of a *lingua democratica.* According to Delanty and Strydom, social constructionism (or constructivism, as it is also called) can be considered as part of a 'turn to context' within the social sciences in which "the demarcation of science from society, the basis of positivism, becomes increasingly difficult." In their eyes this weakening of the demarcation between science and society leads to the conclusion that truth is relative:

> ..the truths of science can no longer be seen as timeless, transhistorical and universal but relative to the scientific context and to the historical location of science....The historical, linguistic and cultural, and more broadly the reflexive, turn in the philosophy of the social sciences has introduced an unavoidable degree of epistemological indeterminacy and uncertainty into the core of science...public debate has to be allowed to play a much greater role in society today.[12]

As Erica Haimes (to whom we owe this reference) underlines chapter 2 of this book, researchers in this field increasingly focus on the normative implications of social constructionism. Referring to the work of D. Hess she states that constructivists are developing into 'post-constructivists': they are interested "not only in how knowledge and technology are constructed but also how they might be *better* constructed, providing alternatives and actively intervening in policy debates."

According to Haimes, we can see that such a suggestion is taking social science through a 'normative turn'. An important contribution to this normative turn is provided by the distinction between two modes of knowledge production as introduced by Michael Gibbons et al. in their book *The New Production of Knowledge.*[13] The well known central thesis of their book concerns the emergence of a new mode of knowledge production, which (speaking with Ulanowicz) builds upon the dominant form of scientific knowledge embodied in the

---

[12] Delanty, 2003, p. 368

[13] Gibbons et al., 1994

Newtonian and Darwinian paradigms, but adds a new element to it, resulting in a new form of knowledge production: Mode 2. Whereas Mode 1 knowledge production is oriented towards questions and problems flowing mainly from internal theoretical and experimental developments within specific *academic* disciplines, such as physics, chemistry and biology and their many sub-disciplines, Mode 2 knowledge production is oriented toward the solution of practical problems, such as the development of new medicines, faster computer chips, new weapon systems or more sustainable forms of energy-production. Thus Mode 1 science is mono-disciplinary and governed by the methodological standards of specific disciplines as defined by handbooks and by reviewing boards of leading journals. In contrast, Mode 2 science is characterized by its inter- and trans-disciplinary character and by the dynamic and heterogeneous character of the research communities involved. Moreover it is 'monitored' by a multiplicity of different stakeholders who define the adequacy of possible solutions for the practical problems at stake from different perspectives, involving diverging interests and values. According to Gibbons et al., during the last decades, the majority of scientific work has shifted from Mode 1 to Mode 2: it is oriented towards the solution of practical problems, has an inter- and transdiciplinary character and is characterized by mixed forms of quality control which in many cases reflect conflicting interests and values. In their eyes, these conflicts provoke new forms of social reflexivity regarding the development of science and technology.[14] This distinction between different of modes of knowledge production helped us in several respects to further clarify the notion of a lingua democratica for genomics.

To start with, distinguishing different forms of producing scientific knowledge offers the opportunity to highlight both the importance and the limitations of social constructionism, as analysed below in more detail by Erica Haimes. Over and against radical forms of social constructionism and their denial of the objective character of scientific knowledge, distinguishing between Mode 1 and Mode 2 allows one to respect deep seated convictions of the majority of practicing scientists and technologists with regard to the objective character of scientific knowledge, without endorsing the opposition

---

[14] Nowotny, H., Scott, P., Gibbons, M., 2002

between objective knowledge on the one hand and subjective opinions and values on the other hand. With the help of the distinction between different modes of knowledge, what was traditionally designated as the objectivity of scientific knowledge can now be recast as the *relative autonomy* of specific forms of Mode 1 research. A practical example taken from the emerging domain of bio semiotics can illustrate this shift.

In a contribution to the Journal *Naturwissenschaft*, one of the central spokesmen of biosemiotics, Marcello Barbieri, tries to make plausible the objective character of organic information as an essential component of molecular reproduction processes. According to Barbieri "...organic information is an objective entity because it is defined by the same sequence for any number of observers."[15] Reformulated in terms of the characteristics of Mode 1 knowledge, the reference to *any* number of observers should be replaced by the members of the *specific* scientific communities involved. The observations involved ask for highly specific (and demanding) forms of socialization, and involve the competent use of complex concepts and instruments. In this specific sense social constructionists (and Thomas Kuhn before them) are certainly right: Mode 1 science asks for the on-going social construction of competent observers. This is an important part of the daily activities of scientists: designing courses, writing handbooks, supervising PhD theses etcetera, thus securing the on-going social reproduction of their discipline and its knowledge base. The transfer of this knowledge base not only involves theoretical, methodological and experimental knowledge, but also the competent use of orienting *metaphors,* for instance the influential metaphor of the genome as a 'blueprint'.[16]

Although social constructivists rightly stress this on-going construction of 'competent observers', they tend to disregard the fact that the relative social closure characterizing Mode 1 has been going on more or less successfully for the past four centuries. Seen from a broader sociological perspective, we are confronted with a quite robust social arrangement, allowing for the development of specific forms of insight and control which are consistently highly valued and for which no functional equivalents are available to date. Seen from

---

[15] Barbieri, 2008

[16] Weele, van der, 2010; see also her contribution below (Chapter 5)

this perspective, it is indeed the relative closure to 'external social influences', combined with an (equally relative) internal openness for radical innovations (such as Ulanowicz's 'third window'), which was traditionally articulated in terms of the objectivity of scientific knowledge. This articulation can now be expressed more adequately as the relative autonomy of Mode 1 science.

However, Mode 1 is only half of the story. It is precisely the distinction between two different modes of producing scientific knowledge which affords the possibility to respect on the one hand deep seated convictions of 'objectivity-oriented, or 'mode 1 oriented' scientists, as central partners to the debates on genomics research, but on the other hand also take into account the 'reflexive turn' in the philosophy and sociology of science. To a certain extent this turn reflects the new forms of entwinement between the acquisition of scientific knowledge and economic, political and social priorities taking shape in the last decades of the twentieth century. This arrangement no longer took the shape of massive investments in 'fundamental research' in the expectation of practically relevant 'spin-off' or the emergence of new possibilities for the practical *application* of scientific knowledge acquired according to the basic pattern of Mode 1. Instead 'external' influences became increasingly *internalized* on the level of the *content* of scientific knowledge. In an interesting reappraisal of their 1994 book, published in 2003, Nowotny, Scott and Gibbons describe this new, Mode 2 entwinement as follows:

> The nature of the research process is being transformed, and this transformation has many separate elements. Scholars disagree about their respective novelty and intensity. However, three trends are generally accepted to be significant – (a) the 'steering' of research priorities, (b) the commercialization of research, and (c) the accountability of science. ... As a result of these and other trends, the research that is variously described as 'pure', 'blue-skies', fundamental, or disinterested, is now a minority preoccupation – even in universities. In Britain, Research Councils and RAE panels now include 'user' representatives alongside more traditional scientific peers. Detailed impact studies and lengthy evaluations have become routine. 'Knowledge' is now regarded not as a public good, but rather as 'intellectual property', which is produced, accumulated, and traded like other goods and

> services in the Knowledge Society. In the process, a new language has been invented – a language of application, relevance, contextualization, reach-out, technology transfer, and knowledge management.[17]

In our eyes this description fits the development of the life sciences and of genomics research very well. It also underlines the urgency of developing a 'lingua democratica for genomics', and for the life sciences in general. In a sequel book to *The New Production of Knowledge,* which appeared in 2003 under the title *Re-Thinking Science*, Nowotny et al. introduced the metaphor of the *agora* to address the question of the democratization of public debates on scientific and technological development:

> This archaism was deliberately chosen to embrace the political arena and the market place – and to go beyond both. The *agora* is the problem-generating and problem-solving environment in which the contextualization of knowledge production takes place. It is populated not only by arrays of competing 'experts', and the organizations and institutions through which knowledge is generated and traded, but also by variously jostling 'publics'. It is not simply a political or commercial arena in which research priorities are identified and funded, nor an arena in which research findings are disseminated, traded, and used. The *agora* is a domain of primary knowledge production – through which people enter the research process, and where 'Mode 2' knowledge is embodied in people and projects. The role of controversies in realizing scientific potential is also played out in the *agora.*[18]

The idea of a *lingua democratica* can be considered as a refinement of this wider notion of *agora,* as a *democratic* environment for the development of Mode 2 science. In contrast to the 'forum of science', where only qualified scientists are allowed to speak,[19] this *agora* should

---

[17] Nowotny, Scott,Gibbons, 2002, pp. 184-185
[18] Nowotny, Scott, Gibbons, 2003, p. 192
[19] Groot, de, 1982

provide a space for democratic deliberation where all stake holders involved can present their own perspective and concerns.[20]

Ad c) The de-centering of politics

In our search for conceptual resources which could help us to develop a shared vocabulary with regard to the possibilities and difficulties involved in the furthering of such democratic environments, we also profited from recent developments in the domain of social and political theory, focusing on far-reaching changes in the structure and internal development of 'post-industrial' societies. Two changes were of special importance for us. In the first place the diminishing importance of 'traditional' democratic organs, such as political parties and parliaments, and the gradual shift of the political centre of gravity to complex and often opaque networks of different social actors. Within these networks the political agenda is constantly being defined and redefined in dynamic interaction between the traditional political centres on one hand and a broad array of social actors on the other hand, such as corporations, NGO's, the mass media and all kinds of divergent and conflicting interest-groups. In the second place, the political priorities emerging from these complex 'local' interactions are structured and influenced on an unprecedented scale by *global* economic and technological developments and shifting geo-political force fields which manifest themselves on the level of national politics, but can hardly be steered on this level, as argued among others by Ulrich Beck and Manuel Castells.[21]

In our eyes this 'de-centering of politics' involves both threats and opportunities for democratic societies. On the one hand they imply a severe limitation of the capacity of the state to function as a centre of steering and unification. Thus the influence and importance of traditional democratic organs such as political parties and the parliament is diminished also. As a consequence, decisions with regard to issues of central social and political importance increasingly tend to come about under the influence of changing coalitions of social actors trying to optimize partial interests in a dynamic and

---

[20] In this respect the ideas developed by Nowotny c.s. show a close resemblance with the analysis of 'postnormal science' by Funtowicz and Ravetz (1991); see also Van der Sluijs, 2012

[21] Beck, 1998, 1999; Castells, 1996, 1997, 1998

opaque network of shifting power relations against the backdrop of global economic and geo-political shifts which none of the actors involved can control. On the other hand this de-centering of politics could also offer opportunities for a democratic renewal of political processes. Within contemporary social and political theory these opportunities are elaborated among others by the proponents of deliberative models of democracy.[22] These models focus on a plurality of modes of association 'in which all affected can have the right to articulate their point of view'. As Seyla Benhabib says:

> It is through the interlocking net of these multiple forms of associations, networks and organizations that an anonymous 'public conversation' results. It is central to the model of deliberative democracy that it privileges such a public sphere of mutually interlocking and overlapping networks and associations of deliberation, contestation and argumentation.[23]

Our research group took this deliberative model of democracy as an important conceptual resource for the clarification and practical realization of a lingua democratica on genomics, but tried to refine and modify this deliberative model in several ways. In the first place we highlight not only the necessity of a public articulation of the different perspectives on the development of genomics, but also underline the importance of taking seriously the *conflicts of interest* between the parties involved and the concomitant tendency to *close off* one's own perspective from open interaction with other perspectives, and impede in this way critical discussion and public debate. In the second place we try to concretize this double focus on public deliberation and on conceptual closure by focusing on the *languages* and core-concepts of central parties to the debates on genomics. By focusing on 'open' versions of the core ideas and central concepts of these different parties, we try to concretize the development of 'interlocking and overlapping networks of deliberation, contestation and argumentation' with regard to genomics. Last but not least, in view of the important role of moral values and of secular and religious worldviews in public debates on genomics, we try to do justice to the important role which 'moral

[22] Benhabib, 1996, 2004; Habermas, 1984, 1985; Honneth, 1996, 2007
[23] Benhabib, 1996, p. 73/74

languages' could play in the *agora* surrounding genomics as a form of Mode 2 research.

Seen from the perspective of the scientists and technologists involved in this *agora,* this last point touches upon one of the most formidable obstacles for the development of a lingua democratica for the life sciences. It is probably safe to say that most scientists and technologists have adjusted, more or less grudgingly, to the fact that in present day post-industrial societies knowledge is no longer "regarded as a public good, but rather as 'intellectual property', which is produced, accumulated, and traded like other goods and services in the Knowledge Society", as Nowotny et al. say. It is also becoming accepted practice that NGO's – considered as representatives of a 'consumer perspective' on new 'knowledge products' - actively participate in institutions and committees influencing the priorities and goals of publicly funded research programmes in the life sciences. Alle Bruggink's analysis (in chapter 10) of the close co-operation which developed in the Netherlands between universities and big corporations in the field of genomics, clearly demonstrates the (partial!) inclusion of the 'consumer perspective' in their deliberations. It seems as if nowadays, universities, corporations and NGO's all fluently speak the language of producing and monitoring new knowledge and 'consuming' (more or less critically) the products flowing therefrom. However the inclusion of moral values and worldviews in these deliberations meets with much greater obstacles. Modern science on the one hand and the moral values embodied in secular and religious worldviews on the other hand traditionally represent *opposing*, mutually *exclusive* perspectives on the world at large. The banning of stem cell research in the United States under the influence of conservative Christian interest groups and the deep frustrations engendered by this ban at the side of the majority of the scientists involved, provide a vivid illustration of the undiminished importance of this opposition.

Seen from the perspective of a lingua democratica however, moral values and worldviews have to be included in the deliberations, because otherwise their unavoidable influence will be shielded from critical examination and discussion. In domains such as food, health, procreation and the genetic enhancement of personal abilities, the 'end users' not only consume the advanced technological products resulting from genomics research, but also the moral values implicitly embodied in those products and in the social practices in which their 'consumption' is embedded. Clear examples of the crucial role of such moral values are provided by the debates on the genetic enhancement of life and the prolongation of the human life span, as advocated among others by 'transhumanists'. Three chapters in part two of this book – by Fernando Suarez, Tatjana Kochetkova and Peter Derkx – are devoted to these discussions.

Moral values and worldviews can thus not be excluded from a lingua democratica for the life sciences. Not only on factual grounds - because their influence is unavoidable – but also because under certain conditions their contribution can *enrich* the deliberation on the development and practical implementation of genomics research. A lingua democratica for the life sciences in general and for genomics in particular has to do justice to the divergent perspectives and conflicts of interest of the different stakeholders involved. The closure which these conflicts tend to engender can be remedied to a certain extent, by strengthening and connecting open and dialogic versions of the different perspectives involved. The development of a lingua democratica depends upon inner connections between the different levels on which the dialogical potential of language manifests itself:

- on the level of academic research and the languages of logic, mathematics, experimentation and analysis
- on the level of public debate and deliberative language in political institutions and in civil societies
- on the level of narrative articulations of moral values and moral dilemma's in literature, art, philosophy, (secular and religious) worldviews and individual life histories.

The idea of a lingua democratica for the life sciences thus points to an extension of the deliberative notion of democracy, as developed by Jürgen Habermas, Seyla Benhabib and others. The dialogical and mediating potential of language manifests itself not only in scientific discourse and in the realm of public debate, but also in the inspirational potential of narrative articulations of moral values and moral dilemma's. In chapter 14 Marlies Huijer illustrates this potential by way of a detailed analysis of the public discussion in the Netherlands on Embryo selection for hereditary breast cancer (Pre-implantation Genetic Diagnosis). Huijer shows that the political decision making process on PGD in the Netherlands was clearly influenced by the impact of personal stories from women with hereditary breast cancer: "The Dutch debate on the allowance of PGD for hereditary breast cancer shows that when individuals directly involved bring their unique life stories into the public, the fierce discussion can change into a more relaxed communication".

The structure of this book

Against this background, we can now sketch the genesis and contents of this book. It contains contributions of all the members of our research group, augmented with papers written by scholars who are working in the same direction. Against the background of the general framework sketched in this introduction, we have grouped the papers in three sections.

The *first* section focuses on the contribution of complexity thinking, of social constructionism and of critical hermeneutics to the development of a *lingua democratica* for genomics. In the opening chapter, *Paul Cilliers* provides a general introduction to critical complexity thinking. In his view we are facing a crisis of knowledge, due to the complexity of the world we are living in and the inadequacy of established scientific and managerial frameworks to adequately deal with this complexity. According to Cilliers "we have to recognise that instrumental rationality, experimental science, social engineering, strategic management and managerialism in general are all strategies which tend to reduce the complexity of the phenomena they deal with, without acknowledging the effects of this reduction." In his view, the result of this acknowledgement should be a radically *critical* position: "Only from such a position can one start engaging with the problems of the world, not in an arrogant way, but with responsibility and a certain modesty". Thus the critical perspective on complexity developed by Cilliers differs in important respects from 'main stream' complexity theory, based primarily on the work of the Santa Fe school. One of the most important differences concerns the necessity highlighted by Cilliers to make *ethical* choices when modelling complex systems. In the concluding paragraph of his contribution he formulates different versions of a complexity-informed 'provisional ethical imperative', in critical reference to Kant's 'categorical imperative'. One of these versions could serve very well as a motto for the further development of a lingua democratic for genomics: "your actions should show a fundamental respect for difference, even as those actions reduce it."

In the second chapter *Erica Haimes* sketches the main features of constructionism, the most influential conceptual paradigm within the domain of social studies of science. Haimes focuses on the question "what social constructionism (and its variations) can contribute to the study of the social, ethical and political aspects of genomics in particular and of the life sciences in general." Social constructionism is rooted in a more general epistemological movement within the social sciences during the last three decades, which points in the same direction as critical complexity thinking, as advocated by Cilliers and others. This movement highlights the historical situatedness of knowledge and the concomitant epistemological indeterminacy and uncertainty. Viewed from this perspective "scientific knowledge suffers the same fate that has fallen on most, if not all, normative discourses…and as Jürgen Habermas insists, the experts have been drawn into legitimation controversies in which publicly acceptable reasons prevail rather than authoritative knowledge".[24] Against this background she then provides a lucid sketch both of the development and main concerns of social constructionism and of important criticisms of constructionism. She concludes with a thesis that resonates in certain respects with Cilliers analysis of the ethical implications of modelling complex systems. This thesis concerns the critical potential and wider political significance of social constructionism. According to Haimes for many scholars in the field the "political power of constructionist-style work is one of its main attractions." This political power rests on two main pillars. On the one hand it rests on the possibilities it offers to 'unmask' the practice of politics itself. Haimes illustrates this critical potential by way of an analysis of parliamentary debates in the UK, more specifically the narrative construction during these debates "of the 'women egg sharer' for somatic cell nuclear transfer research and the deliberate entangling of this social category with that of the 'hybrid embryo' as a strategy to persuade fellow parliamentarians of the necessity for hybrid embryos to 'save' the egg sharer from distress". On the other hand the political significance of social constructionism rests according to Haimes on the possibilities it offers to widen the circle

[24] Delanty and Strydom, 2003, cited by Haimes, below

of those potentially contributing to the lingua democratica by "expanding our awareness of the views and opinions of those beyond the usual stakeholders ... and encourage dialogue between a polyphony of voices."

In the third chapter, *Cees Pieters* follows up the lead provided by Cilliers by way of a complexity-informed reflection on possibilities to enhance the communication between the techno-sciences and the humanities. Pieters argues that theory-bias has obscured for a long time "the important role that *techne* played in the evolution of 'natural philosophy' into contemporary science". The technical character characterizing scientific disciplines such as genomics ensure the possibility of confirmation of theories in continuous cycles of evaluation and testing against something that is 'outside the system', as Pieters says. In contrast, theoretical and ethical perspectives in the humanities show a plurality of differences and a concomitant 'combinatorial explosion of relationships', which makes it very hard for interested outsiders from the techno-sciences to forge meaningful connections to these perspectives. This barrier to the development of a lingua democratica is made even stronger by the apparent lack of interest for the actual practice of techno-scientist *as perceived and experienced by themselves,* manifest in most philosophical and literary criticisms of contemporary techno-science and the dangers supposedly connected with its continuing progress. Against this background Pieters proposed the development of a "philosophy of choice which moves from specific knowledge of a certain domain towards the wider (e.g. philosophical) issues that are related to that domain *through* the language of that domain."

In conclusion of the first part, *Harry Kunneman* tries to develop a conceptual bridge between the foregoing chapters and the questions of value, meaning and politics taking central stage in the remainder of this book. He argues that the development of a lingua democratica for the life sciences, asks for a specific type of knowledge and insight, which he designates as 'Mode 3 knowledge'. Building on the idea of ethical complexity articulated in critical complexity thinking, and on insights from the hermeneutic philosopher Paul Ricoeur, he argues that ethical complexity could serve as a criterion for judging ethical and moral narratives articulating who we are and who we might be in relation to scientific and technological developments in the (life) sciences.

Against this background, the *second* part of this book then tries to answer the question how the democratic character of the public debate on genomics can be improved on the basis of philosophical, ethical and spiritual perspectives with regard to 'hot issues' in these debates, especially the genetic enhancement of human beings, the pre-implantation genetic diagnosis of embryos, and priorities in biotechnological research.

In chapter 5 *Cor van der Weele* looks at the patterns of emphasis and the blind spots that have developed in ethical reflection on genetics and genomics. She argues that different views of the relation between science and society lead to different ways of framing a moral agenda. Very often a new technology like genomics is the starting point for ethical reflection. What will be the moral possibilities and risks of the new technology? In this way the new technology is almost taken for granted and ethical reflection is a set of afterthoughts. Van der Weele argues that different ways of framing ethical issues concerning new technologies should be used to discover the blind spots of each of them. Besides this she argues that a *global* ethics framing deserves more prominence. Ethicists should ask themselves more often what are the most urgent global moral issues and then think about how these can be solved and what the contribution of existing or emerging technologies can be. Starting with the newest technology and see what it can do might be a natural inclination, but it probably is not the most sensible and efficient one from a moral and social point of view.

In chapter 6 *Andries De Smet, Sigrid Sterckx* and *Julian Cockburn* illustrate the importance of such a global perspective in their analysis of a concrete proposal to address the lack of access to medicines in poor countries: the Health Impact Fund (HIF). The HIF provides a interesting example, because it aims to achieve ethical goals using the mechanisms that operate within the existing economic order. One might say that the HIF is a proposal that tries to open up different interests and languages towards each other. The proposal, if it would work as envisaged, would enable both interests to be served at the same time. It tries to create a win-win situation.

In chapter 7 *Fernando Suárez-Müller* develops a philosophical argument against trans-humanism and the idea of corporal immortality from the perspective of humanistic ethics. According to Suarez, human beings do not need superhuman brains or unlimited longevity in order to make wisdom, moral excellence or spiritual fulfilment possible. These are possible already. Human rights, moreover, are based on our common human condition. Making human nature an object of individual choice will undermine our human dignity, the basis for human rights. Just as democracy cannot function without agreeing on a constitution, on some basic values, Suárez-Müller concludes, a *lingua democratica* is not going to function without a *lingua ethica.* Thus in his view, transhumanist improvements to our body are neither necessary nor desirable.

In chapter 8 *Tatjana Kochetkova* defends the importance of transhumanism on the basis of a broad array of spiritual perspectives that share the idea of immortality as the ultimate horizon of human development. Against all kinds of objections from secular and religious quarters, she argues that society should take an open attitude towards the ambitions of the transhumanists to improve the human body. She warns however that transhumanists often forget that this is only half of the story. According to Kotchetkova, an improved immortal body is only a better condition for what really counts from a spiritual perspective: infinite spiritual perfection. In her view this implies corporeal immortality. Se thus concludes that "although transhumanists and spiritual quest supporters climb the mountain from different sides, it is clear that they belong to the same alpinist team."

This discussion of the importance and limitations of the perspective of transhumanists is concluded by *Peter Derkx* in chapter 9. He focuses on one of the main forms of transhumanist enhancement for which genomics is often said to be of crucial importance: substantial life-extension. After sketching the biotechnological possibilities and expectations he summarizes the main arguments originating from ethics (our 'basic values') and secular and religious worldviews. One of his conclusions is that fighting longevity inequalities - within and between countries - deserves strong (though not absolute) moral and political priority. Children born in some countries and social groups may expect far fewer years than people in other regions and social classes. Reducing health and longevity inequalities, however, is notoriously difficult, because powerful economic interests keep them as they are and even increase them.

The *third* section focuses on the role of governments and corporations in determining the direction and content of genomics research and the commercial use of its results and on possibilities to enhance democratic deliberative processes with regard to genomics research. In chapter 10 Alle Bruggink, offers an inside-analysis of the intensive cooperation between universities and corporations in the Netherlands which has developed during the last decades, and the limited role of NGO's in this development. According to Bruggink, "critical comments of NGO representatives, having not been involved in the risks and efforts of the research, are met with mixed feelings." In his view, a more satisfying position of NGO's could be reached by "i) continuing communication between researchers and the NGO, ii) acceptance by the researchers that NGO's can only play their role in the development chain when applicable results are available, and iii) an equal level of scientific quality in presenting the research results and the critical comments of the NGO." He concludes that a lingua democratica "is still rather remote in setting up national research programs but meaningful developments are underway."

In chapter 11 bio-ethicist *Anders Nordgren* provides a vivid illustration of the remoteness of a lingua democratica in the domain of a commercial application of genomics research: online "direct-to-consumer" DNA testing. Nordgren gives a critical analysis of one the strategies used by genomics companies, namely the appeal to *empowerment* of consumers by way of DNA-testing. On the basis of a detailed analysis of the 'information' offered on websites, he concludes that the notion of empowerment can be understood in different ways: "..this ambiguity in combination with its positive connotation makes it very useful in the rhetoric of the consumer genomics companies." According to Nordgren, these claims to empowerment have to be met with great scepticism in view of the risk of misinformation and misinterpretation.

In chapter 12 *Michiel Korthals* develops a philosophical reflection on the prospects of deliberative forms of democracy starting from the different forms of uncertainty connected with genomics research, especially in the field of nutri-genomics. Korthals starts with an analysis of five different types of uncertainty inherently connected in his view with the products of nutri-genomics and then proceeds to argue that deliberative approaches to democracy are particularly well suited to deal with these uncertainties. In his view the currently dominating forms of stakeholder analysis "do not push for transcending the interests and values of the stakeholders involved.

On the contrary, they glue them to their interests and values ... and try to make the best of these." According to Korthals, the approach to deliberative democracy originating with Dewey and enriched with insights from present day thinkers such as Jürgen Habermas and Sheila Benhabib, does offer such incentives for transcending frozen positions. It can even be used to tackle thorny questions at higher, international levels of aggregation: "The deliberative democracy approach to international issues does not strive for explanations, but for illuminating connections, new and promising strategies, and mechanisms that stimulate cooperation between actors involved in forming opinions and making decisions with respect to nutri-genomics. It carefully transcends the local circumstances and outlines various general lines of tackling the interaction between local, regional, national and international. The emergence of international oriented, although fragmented civil societies is an anchor point for those connections. Very often persons or circles of persons act on several levels and these connecting links can help to create channels of understanding."

In chapter 13, *Guido Ruivenkamp* and *Joost Jongerden* introduce a new angle in the debate on the development of a lingua democratica for the life sciences. Instead of concentrating on the quality and more or less inclusive character of debates and discussions, they focus on the political content *embodied* in actual technologies, for instance "new enzymes, which enable farmers to adapt the agricultural produce to those globally organized food conservation methods which reinforce the uncoupling of agricultural produce from local food consumption." Ruivenkamp and Jongerden argue that such technologies can be reconstructed in order to change the 'embodied political content of the biotechnological products.' As a concrete example of such a change, they refer to initiatives of national research institutions in Ghana who cooperate with farmer groups in developing open pollinated enriched maize varieties, which can be cultivated by the farmers in the framework of school feeding programs. In their eyes, this is just one example "illustrating that a multitude of oppressed voices and dialect are appearing and express their quest for another embodiment of politics in key biotechnological products." In combination with open source knowledge programs and with 'Open Forums on technology', such forms of embodying more democratic political arrangements in biotechnology itself, could considerably strengthen the development of a lingua democratica for the life sciences.

In the concluding chapter 14, *Marli Huijer* presents an analysis of the public discussion in the Netherlands on Embryo selection for hereditary breast cancer (Pre-implantation Genetic Diagnosis). Huijer shows how the rather closed positions of political parties and politicians on pre-implantation genetic diagnosis (PGD) can be opened up when in the public media individual citizens tell their personal stories about living with hereditary breast cancer. By foregrounding the political importance of story telling, Huijer's analysis points to a broader notion of deliberative democracy in which arguments, stories and emotions can be connected with each other on an equal footing.

* We would like to thank Erica Haimes for her valuable comments on an earlier version of this introduction.

Literature

Barbieri, M. (2003). *The Organic Codes: an Introduction to Semantic Biology.* Cambridge: Cambridge University Press.

Barbieri, M. (2008). "Biosemiotics: a new understanding of life". *Naturwissenschaften, 95*, 577-599.

Beck, U. (1998). *World Risk Society.* Cambridge: Polity Press.

Beck, U. (1999). *What is Globalization?.* Cambridge: Polity Press.

Benhabib, S. (1996). *Democracy and Difference.* Princeton: Princeton University Press.

Benhabib, S. (2004). *The Rights of Others.* Cambridge: Cambridge University Press.

Broad, C.D.,(1930). *Five Types of Ethical Theory.* Harcourt, Brace and Co., New York

Castells, M. (1996). The Rise of the Network Society, The Information Age: Economy, Society and Culture Vol. I. Cambridge, MA: Blackwell.

Castells, M. (1997). The Power of Identity, The Information Age: Economy, Society and Culture Vol. II. Cambridge, MA: Blackwell.

Castells, M. (1998). End of Millennium, The Information Age: Economy, Society and Culture Vol. III. Cambridge, MA: Blackwell.

Cilliers, P. (1998). Complexity and postmodernism : understanding complex systems. New York: Routledge.

Cilliers, P. (2005), "Complexity, Deconstruction and Relativism". *Theory, Culture and Society, 22*(5), 255-267.

Delanty, G. S., G. (Ed.), (2003). Philosophies of Social Science: The Classic and Contemporary Readings. Maidenhead: Open University Press.

Gibbons, M., Limoges, C., Nowotny, H., Schwartzmann, S., Scott, P., & Trow, M. (1994). *The New Production of Knowledge. The Dynamics of Science and Research in Contemporary Societies.* London: Sage Publications.

Groot, A. d. (1982). *Academie en forum: over hoger onderwijs en wetenschap.* Meppel: Boom.

Habermas, J. (1984). *The Theory of Communicative Action, Vol. 1: Reason and the Rationalization of Society.* Boston: Beacon Press.

Habermas, J. (1985). *The Theory of Communicative Action, Volume 2: Lifeworld and System: A Critique of Functionalist Reason.* Boston, Massachusetts: Beacon Press.

Hoffmeyer, J. (2009). *Biosemiotics. An examination into the Signs of Life and the Life of Signs.* Scranton and London: University of Scranton Press.

Holland, J. H. (1998). *Emergence : from chaos to order.* Reading, Mass.: Addison-Wesley.

Honneth, A. (1996). *The Struggle for Recognition: The Moral Grammar of Social Conflicts.* London: Polity Press.

Honneth, A. (2007). *Disrespect: The Normative Foundations of Critical Theory.* London: Polity Press.

Irigaray, L. (1993). *An Ethics of Sexual Difference.* Ithaca: Cornell University Press.

Kaufmann, S. (1993). *Self-Organization and Selection in Evolution.* Oxford: Oxford University Press.

Kaufmann, S. (2000). *Investigations.* Oxford: Oxford University Press.

Markos, A. (2002). *Readers of the Book of Life: Contextualizing Developmental and Evolutionary Biology.* Oxford: Oxford University Press.

NGI. (2007). *Report of the NGI [Netherlands Genomics Initiative] 2002-2007.* Retrieved from www.genomics.nl.

Nowotny, H., Scott. P. , Gibbons, M. . (2002). *Re-Thinking Science. Knowledge and the Public in an Age of Uncertainty.* Cambridge: Polity Press.

Nowotny, H., Scott. P., Gibbons, M. (2003). "Introduction: Mode 2 Revisited: The New Production of Knowledge", *Minerva, 41*, 179-194.

Popper, K. (1990). *A World of Propensities.* Bristol, UK: Thoemmes.

Sebeok, T. A. (1992). *Biosemiotics: The Semiotic Web.* Berlin: Mouton de Gruyter.

Sebeok, T.A. (2001). "Biosemiotics: its Roots, Proliferation and Prospects". *Semiotica, 134* (1/4), 61-78.

Sluijs, J.P. van der (2012) "Uncertainty and Dissent in Climate Risk Assessment: A Post-Normal Perspective", Nature and Culture, Vol. 7, no 2, pp. 174-195

Taylor, C. (1989). *Sources of the self : the making of the modern identity.* Cambridge: Cambridge University Press.

Taylor, C. (1992). *The ethics of authenticity.* Cambridge, MA: Harvard University Press.

Ulanowicz, R. E. (2009). *A Third Window. Natural Life beyond Newton and Darwin.* West Conshohocken, Pennsylvania: Templeton Foundation Press

Weber, M. (1968*), Gesammelte Aufsätze zur Wissenschaftslehre*, Tübingen: Mohr (1922)

Part 1

Complexity Theory, Social Constructionism and Hermeneutics

1

A Crisis of Knowledge: Complexity, Understanding and the Problem of Responsible Action

Paul Cilliers

The time is out of joint

Hamlet

More than any time in history mankind faces a crossroads. One path leads to despair and utter hopelessness, the other to total extinction. Let us pray that we have the wisdom to choose correctly.

Woody Allen

Introduction

Perhaps not everybody has noticed, but we are facing a crisis. Not just in the Middle East or Africa, but everywhere. This crisis is not simply a political or economic one — although it always will be that too. It has to do with the way in which we understand and describe the world in the first place; and how this understanding affects the way in which we make decisions. We are simply in more trouble than we ever thought we were. This claim is not meant to be apocalyptic, nor is it a new one. The crisis has been with us for a long time. Nevertheless, it now has urgency, if only because we are now confronted with it.

This crisis of understanding is primarily a function of the complexity of the world we live in. I will argue that we cannot develop a complete understanding of something complex. The mess we are in is the result of a whole history of philosophical, political and economic strategies which tried to circumvent this problem, and to a large extent, these strategies are still the ones governing our actions. The question before us then is how to deal with a complex world after we have acknowledged that we have a limited understanding of it? I do not have a clear answer to this question. As a matter of fact, the argument will be that a clear answer will return us to exactly the state we are trying to escape from.

If this is the case, what can we do? I will argue that in the first place we have to acknowledge the complexity and the difficulties it entails. This has not been done sufficiently, not even by most complexity theorists. We have to recognise that instrumental rationality, experimental science, social engineering, strategic management and managerialism in general are all strategies which tend to reduce the complexity of the phenomena they deal with, without acknowledging the effects of this reduction. The result of this acknowledgement should be a radically critical position. Only from such a position can one start engaging with the problems of the world, not in an arrogant way, but with responsibility and a certain modesty.

Acknowledging complexity has far-reaching implications for all domains of human endeavour. Three of them, all three extremely broad, are of central concern: the domain of epistemology (what is the status of our knowledge of the world); the ethical/ political/ economic domain (how do we act in the world and why) and the domain of sustainability (how do we maintain a world in which we can be fully human). My focus will be on the first of these domains. Why is it so difficult to know complex things?

Why Reductionism Fails

To a large extent we still live in a world where "scientific" knowledge trumps all other forms of knowledge. This state of affairs is a legacy of a certain interpretation of Enlightenment thinking. In this interpretation, the quest for verifiable knowledge, at least since the Renaissance, presupposes the need for objectivity. Novotny et al. describe this process in the following way: "

> In its historical contest with religion, a triumphant science acquired a monopoly of describing and explaining 'reality', which both resisted and also validated human wishes, fancies and follies. Because the physical world, including

> its chemical and biological processes, came to be regarded as the most substantial component of the 'real world', a scientific definition of reality became ever more plausible. As a result the authority, values and practices of science permeated many other dimensions of society. The everyday world shrank to what scientists had 'discovered' and were able to exploit.[1]

This traditional or, as it is often called, "modernist" style of scientific thinking is no longer adequate – to the extent that it ever was. This is not the result of a frivolous postmodern reaction to modernity , nor is it merely because of some logical problem with the verification of experimental processes (in the tradition of Popper, Kuhn or Feyerabend), but a result of the complexity of the phenomena we deal with. As Novotny et al. state: "Contemporary society is characterized – irreversibly – by pluralism and diversity and also, we argue, volatility and transgressivity. It can no longer be understood either in terms of the norms and practices of scientific rationality …"[2]

What is at stake when we deal with complex things is thus the appropriate style of rationality. The argument is that the traditional modernist rationality is not adequate to cope with complexity. This modernist form of rationality was established in the first half of the 17th century and based on the ideas of Galileo, Newton and Descartes in the context of a more settled Europe after the peace of Westphalia.[3] This style of thinking was fundamental to the establishment of the Royal Society, or more precisely, The Royal Society of London for the Improvement of Natural Knowledge, with the now extremely disconcerting motto *nullius in verba.* Edgar Morin gives specific content to the inadequacy of what he calls'classical science'.[4] For him "classical science rejected complexity in virtue of three fundamental explanatory principles:

1. The principle of universal determinism, illustrated by Laplace's Daemon, capable, thanks to his intelligence and

---

[1] Novotny, Scott and Gibbons, (Eds), 2001, p. 50/51
[2] Novotny, Scott, and Gibbons (Eds), 2001, p. 21
[3] Toulmin, 1990
[4] Morin, 2007, p. 5

extremely developed senses, of not only knowing all past events, but also of predicting all events in the future.

2. The principle of reduction, that consists in knowing any composite from only the knowledge of its basic constituting elements.
3. The principle of disjunction, that consists in isolating and separating cognitive difficulties from one another, leading to the separation between disciplines, which have become hermetic from each other."

For Morin, this tradition has led to wonderful results, but only in a limited context. In order to deal with a complex world, however, we need to acknowledge the limitations of this approach. An epistemological shift is required which replaces 'reduction' with 'distinction' and 'disjunction' with 'conjunction.[5]

In many ways one can argue that we are dealing with an epistemological crisis in contemporary theory. It is generally acknowledged that simple reductive thinking is not adequate, but there is also a fundamental fear that abandoning it may lead to relativism. If this was merely a theoretical problem, we could let the philosophers argue about it. Unfortunately different strategies of thinking lead to different forms of action in the world. The disastrous effects of reductive thinking are evident in many spheres; including the social, the political, the economical and the environmental. The epistemological shift Morin talks about is, therefore, not merely a theoretical issue, but one with practical and ethical implications.

In what follows, an attempt will be made to give some content to the nature of this shift. It will be argued that we do not have a consistent or complete language which can replace the reductive one, but that we can articulate some of the shortcomings and limitations. This articulation is primarily critical, but it is a critique which leads to different forms of action, and thus to real results.

---

[5] Morin, 2007, p. 10

Dealing with Complexity

An interest in complexity science has blossomed in the last three decades or so. Fuelled by the work of, amongst many others, Prigogine, Maturana and Varela, Mandelbrot, Kaufman, Bak, Gell-Mann and a generation of chaos theorists, the characteristics of complex systems have been studied intensively. Numerous institutions have been founded devoted solely to the investigation of these issues. The mere fact that a lot of attention is being paid to complexity is, however, no guarantee that the epistemological shift referred to above has taken place. Morin is quite explicit that, even in complexity theory, the dominance of a traditional instrumental rationality has largely been retained. In order to make this explicit, he makes a distinction between 'restricted complexity' and 'general complexity'.[6]

Restricted complexity is, for Morin, exemplified in those approaches to complexity which developed from chaos theory and fractal mathematics. These approaches focus on underlying patterns and universal principles which are still highly reductive in nature .

> Restricted complexity made [...] possible important advances in formalization, in the possibilities of modelling, which themselves favor interdisciplinary potentialities. But one still remains within the epistemology of classical science. When one searches for the 'laws of complexity', one still attaches complexity as a kind of wagon behind the truth locomotive, that which produces laws. A hybrid was formed between the principles of traditional science and the advances towards its hereafter. Actually, one avoids the fundamental problem of complexity which is epistemological, cognitive, paradigmatic. To some extent, one recognizes complexity, but by decomplexifying it. In this way, the breach is opened, then one tries to clog it: the paradigm of classical science remains, only fissured.[7]

General complexity, Morin argues, is not merely a methodology; it involves a rethink of our fundamental definitions of what knowledge

---

[6] Morin, 2007

[7] Morin, 2007, p. 10. Original translation.

is. When dealing with complexity, the traditional method of analysis does not work. What is more, the divide between subject and object cannot be maintained in any clear way. This is how Morin formulates it:

> In opposition to reduction, [general] complexity requires that one tries to comprehend the relations between the whole and the parts. The knowledge of the parts is not enough, the knowledge of the whole as a whole is not enough … Thus, the principle of reduction is substituted by a principle that conceives the relation of whole-part mutual implication. The principle of disjunction, of separation (between objects, between disciplines, between notions, between subject and object of knowledge), should be substituted by a principle that maintains the distinction, but that tries to establish the relation"[8]. From this formulation it should be clear that Morin is not advocating a relativistic position, nor is he arguing for a "generality" which is naively holistic or vague. One cannot say anything without making distinctions, but these distinctions are always contextualised within a set of relationships.[9]

To my mind, Morin is absolutely correct about the fact that much of current complexity theory remains trapped within a traditional rationality. I witnessed a prominent theorist in the field claiming that everything which happens in society is reflected in the Dow Jones index, thereby reducing all the complexities of human society to a single index, and a financial one at that. The current interest that many complexity theorists have in power laws, is evidence of a similar form of reduction. It rests on the problematic assumption that the world is fractal in nature. At least in the biological and social domains scale matters very much. Our knowledge of things complex cannot be free-floating and abstracted; it is contingent and historically determined. Perhaps certain natural phenomena are more amenable to descriptions in terms of a 'restricted' complexity. But to develop a deeper understanding of ecological, human and social phenomena, we will have to move beyond a simplified Enlightenment rationality, even if that rationality is spruced up with a bit of chaos theory. For

---

[8] Morin, 2007, p. 10/11

[9] Morin, 2007, p. 18/20

that we need to develop an understanding of complexity that is brave enough to think the uncomfortable consequences of its insights through to the end.

Characterising Complexity

A "general" notion of complexity does involve a host of epistemological problems, but that does not mean that one cannot say substantial things about it. According to conventional academic practice it would be appropriate to provide some definition of what is meant by "complexity". Yet, there is something inherently reductionist in the process of definition. This process tries to capture the precise meaning of a concept in terms of its essential properties. It would be self-defeating to start an investigation into the nature of complexity by using exactly those methods we are trying to criticise. On the other hand, we cannot leave the notion of "complexity" merely dangling in the air; we have to give it some content. This will be done by making a number of distinctions that will constrain the meaning of the notion without pinning it down in a final way. The characterisation developed in this way is thus not final – in specific contexts there may be more characteristics one could add, and some of those presented here may not always be applicable – but it helps us to make substantial claims about the nature of complexity, claims that may shift our understanding in radical ways.

In the first place one should recognise that complexity is a characteristic of a system. Complex behaviour arises because of the interaction between the components of a system. One cannot, therefore, focus on individual components, but on their relationships. The properties of the system emerge as a result of these interactions; they are not contained within individual components.

A second important issue is to recognise that a complex system generates new structure internally. It is not reliant on an external designer. This process is referred to, somewhat problematically, as self-organisation. In reaction to the conditions in the environment, the system has to adjust some of its internal structure. In order to survive, or even flourish, the tempo at which these changes take place is vital (see Cilliers, 2007 for detail in this regard). A comprehensive discussion of self-organisation is beyond the scope of this chapter (see Chapter 6 in Cilliers, 1998 for such a discussion), but some aspects of self-organisation will become clear as we proceed.

An important distinction can be made between "complex" and "complicated" systems. Certain systems may be quite intricate, say something like a jumbo jet. Nevertheless, one can take it apart and put it together again. Even if such a system cannot be understood by a single person, it is understandable in principle. Complex systems, on the other hand, come to be in the interaction of the components. If one takes it apart, the emergent properties are destroyed. If one wishes to study such systems, examples of which are the brain, living systems, social systems, ecological systems and socio-ecological systems, one has to investigate the system as such. It is exactly at this point that reductionist methods fail.

One could argue, however, that emergence is a name for those properties we do not fully understand yet. Then complexity is merely a function of our present understanding of the system, not of the system itself. Thus one could distinguish between epistemological complexity – complexity as a function of our description of the system – and ontological complexity – complexity as an inherent characteristic of the system itself. Perhaps, the argument might go, all complexity is merely epistemological, that finally all complex systems are actually just complicated and that we will eventually be able to understand them perfectly.

If one follows an open research strategy — a strategy which is open to new insights as well as to its own limitations — one cannot dismiss the argument above in any final way. Nevertheless, until such time as the emergent properties of a system are fully understood, it is foolish to treat them as if we understand them already. Given the finitude of human understanding, some aspects of a complex system may always be beyond our grasp. This is no reason to give up on our efforts to understand things clearly. It is the role of scientific enquiry to be as exact as possible. However, there are good reasons why we have to be extremely careful about the reach of the scientific claims we make. In order to examine these reasons in more detail, a more systematic discussion of the nature of complex systems is required. The following characteristics will help us to do this:

1. Complex systems are open systems.
2. They operate under conditions not at equilibrium.
3. Complex systems consist of many components. The components themselves are often simple (or can be treated as such).
4. The output of components is a function of their inputs. At least some of these functions must be non-linear.
5. The state of the system is determined by the values of the inputs and outputs.
6. Interactions are defined by actual input-output relationships and these are dynamic (the strength of the interactions change over time).
7. Components, on average, interact with many others. There are often multiple routes possible between components, mediated in different ways.
8. Many sequences of interaction will provide feedback routes, whether long or short.
9. Complex systems display behaviour that results from the interaction between components and not from characteristics inherent to the components themselves. This is sometimes called emergence.
10. Asymmetrical structure (temporal, spatial and functional organisation) is developed, maintained and adapted in complex systems through internal dynamic processes. Structure is maintained even though the components themselves are exchanged or renewed.
11. Complex systems display behaviour over a divergent range of timescales. This is necessary in order for the system to cope with its environment. It must adapt to changes in the environment quickly, but it can only sustain itself if at least part of the system changes at a slower rate than changes in the environment. This part can be seen as the 'memory' of the system.
12. More than one legitimate description of a complex system is possible. Different descriptions will decompose the system in different ways and are not reducible to one another. Different descriptions may also have different degrees of complexity.

If one considers the implications of these characteristics carefully a number of insights and problems arise:

- The structure of a complex system enables it to behave in complex ways. If there is too little structure (i.e. many degrees of freedom), the system can behave more randomly, but not more functionally. The mere 'capacity' of the system (i.e. the total amount of degrees of freedom available if the system was not structured in any way) does not serve as a meaningful indicator of the complexity of the system. Complex behaviour is possible when the behaviour of the system is constrained. On the other hand, a fully constrained system has no capacity for complex behaviour either. This claim is not quite the same as saying that complexity exists somewhere on the edge between order and chaos. A wide range of structured systems display complex behaviour.
- Since different descriptions of a complex system decompose the system in different ways, the knowledge gained by any description is always relative to the perspective from which the description was made. This does not imply that any description is as good as any other. It is merely the result of the fact that only a limited number of characteristics of the system can be taken into account by any specific description. Although there is no a priori procedure for deciding which description is correct, some descriptions will deliver more interesting results than others.
- In describing the macro-behaviour (or emergent behaviour) of the system, not all the micro-features can be taken into account. The description on the macro-level is thus a reduction of complexity, and cannot be an exact description of what the system actually does. Moreover, the emergent properties on the macro-level can influence the micro-activities, a phenomenon sometimes referred to as "top-down causation". Nevertheless, macro-behaviour is not the result of anything else but the micro-activities of the system, keeping in mind that these are not only influenced by their mutual interaction and by top-down effects, but also by the interaction of the system with its environment. When we do science, we usually work with descriptions which operate

mainly on a macro-level. These descriptions will always be approximations of some kind.

These insights have important implications for the knowledge-claims we make when dealing with complex systems. Since we do not have direct access to the complexity itself, our knowledge of such systems is in principle limited. The problematic status of our knowledge of complexity needs to be discussed in a little more detail.

Complexity, Limits and Knowledge

The argument that our understanding of complex systems is problematic in principle can be made in the following way: to fully understand a complex system, we need to understand it in all its complexity. Furthermore, because complex systems are open systems, we need to understand the system's complete environment before we can understand the system, and, of course, the environment is complex in itself. There is no human way of doing this. The knowledge we have of complex systems is based on the models we make of these systems, but in order to function as models – and not merely as a repetition of the system – they have to reduce the complexity of the system. This means that some aspects of the system are always left out of consideration. The problem is compounded by the fact that that which is left out, interacts with the rest of the system in a non-linear way and we cannot, therefore, predict what the effects of our reduction of the complexity will be, especially not as the system and its environment develop and transform in time .

We cannot have complete knowledge of complex systems; we can only have knowledge in terms of a certain framework. There is no stepping outside of complexity (we are finite beings), thus there is no framework for frameworks. We choose our frameworks. This choice need not be arbitrary in any way, but it does mean that the status of the framework (and the framework itself) cannot be used as the basis for objective knowledge. The generation of knowledge of complex systems is an exploratory process. As the context in which this knowledge is to be useful changes, we will have to continually revise the framework which generates this knowledge. Our knowledge of complex systems is thus always provisional. We have to be modest about the claims we make about such knowledge.

An understanding of knowledge as constituted within a complex system of interactions would, on the one hand, deny that knowledge can be seen as atomised "facts" that have objective meaning. Knowledge comes to be in a dynamic network of interactions, a network that does not have distinctive borders. On the other hand, this perspective would also deny that knowledge is something purely subjective, mainly because one cannot conceive of the subject as something prior to the "network of knowledge", but rather as something constituted within that network. The argument from complexity thus wants to move beyond the objective/subjective dichotomy, as Morin (2007) also argues. The dialectical relationship between knowledge and the system within which it is constituted has to be acknowledged. The two do not exist independently, thus making it impossible to first sort out the system (or context), and then to identify the knowledge within the system. This co-determination also means that knowledge, and the system within which it is constituted, is in constant transformation. What appears to be uncontroversial at one point may not remain so for long.

One should also be careful not to interpret this state of affairs as somehow inadequate, as something to be improved upon. There is a necessary relationship between the imposition of a limiting framework and the generation of knowledge. One cannot have knowledge without a framework. Despite the fact that our knowledge is of necessity limited, these limits are enabling, they allow us to make claims which are neither relativistic nor vague (see Cilliers 2005). At the same time, however, such knowledge is not the result of free-floating truths; it is contextualised in time and space. Because it is not objective, and because we know that, we cannot use this knowledge as if it is objective. There is always a normative dimension to the claims we make, and we have to stand in for them. We cannot shift the responsibility for the effects of our claims onto some process we call "scientific".

The Problem of Action

If the central insight from complexity — that our decisions will always have to be based on an incomplete understanding and thus will always have to be provisional — is valid, then it is clearly very difficult to use this insight to guide our actions. We can neither form a complete picture of the current situation and what it means, nor can we formulate an exact goal for which we should aim. We know that we cannot get it right.

Where does this leave us? Does it imply that we can do no more than slump into a dejected resignation? I think not. Yet to generate a more positive framework for action, without violating the insights form complexity, is not easy. Let me make some preliminary remarks.

In the first place, slumping into a dejected resignation is much better than to act with the self-confident arrogance of someone who thinks he knows what is right. There is a certain sense in which we should be knocked speechless by our ignorance, and by the mess we have created by acting on that ignorance. This is the essence of a truly critical position and a vital first step towards responsible action. We do not have to have an alternative before we can realise something is wrong or before we denounce it. I do not have an alternative to free-market economy, but I can see that it is wrong. If you do not recognise that fact, you have neither eyes nor a heart. We should net let ourselves be forced by instrumental rationality, nor by the demands of a performance culture, to relinquish our critique. Nevertheless, this should not lead to inaction.

In the second place we should realize that our decisions and actions do not only have effects on the world and on others, but also on ourselves. We are constituted through our decisions and actions, or more precisely, we mutually constitute each other in a rich, non-linear network of interactions. In a sense, we are simply the result of our decisions. If I choose a position of power and dominance, or a position which values material wealth over aesthetic value, that is simply the kind of person I am. Therefore, when I make a specific decision or perform a specific act, I do not do so from a fresh platform. I do so from a web of previous acts by myself and by those with whom I interact, a web in which I have already established a certain identity. And, at the same time, I am busy constituting and transforming that web. We are not atomistic individuals with a duty to act, however that duty is conceived (rationally, transcendentally or religiously), we are constituting our humanity by the way in which we do battle with what is before us, by the way in which we enter into the "agonistics of the network". "Tomorrow in the battle think on me" says Javier Marias .

The acknowledgement of complexity, thirdly, thus has an effect on our attitude towards action. The lack of complete knowledge does not mean that we should not act, but it does mean that we should do so with modesty.[10] Every decision should be the result of careful and critical reflection (which is why we will always need Philosophy) and should unfold in time, neither too quickly nor too slowly. The principle of modesty does not determine the content of our actions, but it does affect the way in which we make them. Nevertheless, if care and provisionality determine the framework of our actions, it will also affect the content of our actions. Walking through a minefield carefully does not mean that you will not step on one, but you will attend to every detail around you and you will have a better chance than one who just stomps through. Moreover, you will be extremely sensitive to, as well as critical of, any advice others with a different perspective may give.

Taking these considerations into account places one in a difficult position but, as acknowledged from the start, not a uniquely new one. The dilemma has been recognised by critical thinkers repeatedly, Nietzsche and The Frankfurt School being prime examples. One contemporary form of critical philosophy, namely, deconstruction, can be shown to share many of the insights and problems generated by acknowledging complexity. Before returning to the problem of action, this can be examined briefly.

Complexity and Deconstruction

The characterization of complexity developed here, and the subsequent implications for the status of our knowledge of complex matters, resonates in many ways with some of the central insights of

---

[10] Cilliers, 2005

post-structural philosophy, specifically with deconstruction. In *Complexity and Postmodernism*, I elaborate in detail on the structural similarities between a relational and distributed understanding of complexity on the one hand, and the post-structural understanding of textuality and language (in the most general sense of the word) on the other.[11] The central argument is that there is a striking similarity between complex systems, understood as a network of non-linear interactions with lots of feedback paths, and two theoretical positions foundational to structuralism and post-structuralism: the characterization of language as an interrelated system of signs of Ferdinand de Saussure and the neurological model of the brain developed by the early Freud. Both Saussure and Freud were "read" in some detail by Derrida and a transformation of their ideas form a central part of his position. Relating these arguments with complexity theory allows, on the one hand, an enrichment of complexity thinking (by e.g. incorporating notions like différance) and, on the other, a more systematic and rigorous reading of deconstruction.

One should remember that to deconstruct a position is not to dismiss it, but to take it rather seriously. Derrida's deconstruction of Saussure, therefore, acknowledges the importance of his central argument, but shows how he does not follow his argument all the way through. Saussure shows that the meaning of a sign is not a result of an essential characteristic of the sign itself, but of its relationships with other signs. Nevertheless, Saussure believes that by tracing all the relationships in the system of signs, one can eventually reconstruct the correct meaning of the sign. Derrida argues that this is not possible since the very use of a sign already "disturbs" the meaning of the sign, a disturbance which percolates through the network of relationships, also back to the sign itself, thereby altering the meaning of all the elements in the system. The meaning of a sign is thus continually deferred, it cannot be resolved in any final way. This does not mean that meaning does not exist, it always already exists, but it is also constantly transformed.

It is clear that Derrida's argument is based on the fact that meaning is constituted through complex interaction. Although he did not

---

[11] Cilliers, 1998

elaborate on a theory of complexity explicitly, a sensitivity to complexity permeates his thinking. I think that there are still many fruitful insights to be gained from a sustained interaction between deconstruction and complexity theory. This work is still to be done, but some initial insights can be gained from comparing some of Morin's arguments with deconstruction.

The first important insight follows from his description of a "restricted" understanding of complexity. This understanding is clearly related to the Saussurian position. It acknowledges the basic structure of complexity, but balks before the more radical consequences. In Morin's terms, it opens up the understanding towards relational thinking, but it cannot get rid of the reductive apparatus that should qualify this work as 'science'. As a result, this approach to complexity – and I would put most of the work done under the umbrella of the so-called Santa Fè School in this category – reverts to an instrumental strategy in the hope of making purely objective claims in the same way as Saussure's claim that we can get at the correct meaning of the sign. It is precisely this denial of a normative element in our dealing with complexity which makes this position "restricted". In developing a deeper understanding of what a "general" understanding of complexity could be, something for which Morin thinks we do not yet have a language, insights from deconstruction could play a vital role.

One such insight could be the idea of the "double movement". Derrida argues that the strategy of deconstruction involves a "double" activity. In deconstructing a system, one has to make use of the resources provided by the system itself. One is thus simultaneously confirming and undermining central elements of the system. This simultaneous give and take is a much more complex process than simply replacing something with something else. It implies that one transforms something by using the thing itself in novel ways. Deconstruction is thus not a critique from the outside, a critique which knows where it stands and what it wants to do. It is a critique which acknowledges that it is in transformation itself because it cannot depart from a perfect understanding, neither of itself, nor of that which it is transforming.

In a selection of newly translated papers, Morin describes the way in which he thinks we should deal with complexity in very similar terms to that of deconstruction. He argues that when dealing with complexity, we cannot escape contradiction, and that we should not mask this contradiction with a "euphoric vision of the world":

> [The order/disorder/organization relationship] is a typically complex idea in the sense that we have to bring together two notions – order and disorder – that logically seem to exclude each other. In addition, we might think that the complexity of this idea is even more fundamental. ... We arrive by entirely rational means at ideas that carry a fundamental contradiction.[12]

He continues:

> In the classical view, when a contradiction appears in reasoning, it is a sign of error. You have to back up and take a different line of reasoning. However, in a complex view, when one arrives via empirical rational means at contradictions, this points not to an error but rather to the fact that we have reached a deep layer of reality that, precisely because of its depth, cannot be translated into our logic.[13]

The point he wants to emphasize is that we cannot deal with complexity without employing a self-critical rationality, that is, a rationality which makes no claim for objectivity, or for any special status for the grounds from which the claim was made.

> Humanity has two types of madness. One is obviously very visible, it's the madness of absolute incoherence, of onomatopoeia, of words spoken randomly. The other is much less visible: it is the madness of absolute coherence. Against this second madness, the resource is self-critical rationality and recourse to experience[14].

In order to maintain this self-critical rationality, he argues "that there are three principles that can help us to think complexity". The first he calls "dialogic". "The dialogic principle allows us to maintain the duality at the heart of unity. It associates two terms that are at the same time complementary and antagonistic."[15]

---

[12] Morin, 2008, p. 41
[13] Morin, 2008, p. 45
[14] Morin, 2008, p. 48
[15] Morin, 2008, p. 49

The second principle is that of "organised recursion". This principle argues for an understanding which "has broken away from the linear idea of cause and effect, of product/producer or structure/superstructure, because everything that is product comes back on what produces it in a cycle that is itself self-constitutive, self-organizing, and self-producing."[16]
The third is the "holographic principle". This principle argues that the characteristics of a system is distributed, not localised. The activities of the parts and the occurrences on the macro-level participate in producing the system (see above). "The idea of the hologram surpasses both reductionism, which can see only the parts, and holism, which sees only the whole"[17].

These three principles are clearly interlinked. The holographic principle is an effect of the recursive principle which is linked to the dialogic principle. This constellation of ideas thus argues for a kind of double movement, an acknowledgment of the play of différance, very similar to that of deconstruction. There is a coupling between what is being observed and how it is being observed; they are folded into each other. Despite our bravest attempts, we cannot extract ourselves from these folds cleanly. Nevertheless, this is what we do, and, in a contradictory way, have to do when we do science:

> ... every system of thought is open and contains a breach, a gap in the opening itself. But we have the possibility to hold meta-points of view. The meta-point of view is only possible if the observer-conceiver integrates himself or herself into the observation and the conception. This is why complex thought requires the integration of the observer and the conceiver in its observation and conception[18].

The kind of understanding of complexity proposed here certainly does not produce a clear "method" which can be followed in any automatic way. Morin is also clear on this: "I can't pretend to pull a paradigm of complexity out of my pocket."[19] Nevertheless, one can

---

[16] Morin, 2008, p. 49/50
[17] Morin, 2008, p. 50
[18] Morin, 2008, p. 51
[19] Ibidem.

announce the implications of this position without proclaiming a new orthodoxy. In his words, "one can be the Saint John the Baptist of the paradigm of complexity and announce its coming without being its Messiah."[20]

The kind of language used here clearly indicates the presence of a normative dimension, something ethical, in our attempts to think and act when confronted with complexity. Perhaps something can be said about this "ethics of complexity".

The Provisional Imperative

The argument for the inevitability of an ethical position when dealing with complexity can be made in the following way: Since we cannot have complete knowledge of complex things we cannot "calculate" their behaviour in any deterministic way. We have to interpret and evaluate. Our decisions always involve an element of choice which cannot be justified objectively. What is more, no matter how careful our actions are considered, they may turn out to have been a mistake. Thus, acknowledging that values and choice are involved does not provide any guarantee that good will come of what we do. Complexity tells us that ethics will be involved, but does not tell as what that ethics actually entails. The ethics of complexity is thus radically or perpetually ethical. There is no a priori principle we can follow nor utility we can compute. We do not escape the realm of choice.

It is clear that, beyond the realisation that we are always in trouble, this position does not generate any substantial guidelines. It constitutes a radically critical position, the main component of which is self-critique. The question is, can it be made more substantial? A first response would be that it is better to make the value judgements explicit than to claim a false objectivity. In this way the complexity of the problem can be opened up and the differences respected. But perhaps the critical position itself constitutes a kind of ethical strategy, similar to Kant's categorical imperative.

The logic by which Kant deduces the categorical imperative can roughly be described as follows : for a moral principle to be

[20] Morin, 2008, p. 52

universally valid, it has to be purely abstract and formal. It cannot be constructed empirically, or take contingent matters into account. The only rule which conforms to this specification is has to be something like "follow this rule". But since the moral rule Kant is looking for has to be universally valid, he can reformulate this abstract rule to something like "follow only rules which are universal", or "follow only those rules which you would always also want others to follow". By combining the purely formal principle with the notion of universalisability, Kant can generate a formulation which actually does say something about ethics.

There are many problems with the categorical imperative (see e.g. Kant/Paton 1948).The main one is the result of the very attempt to be universal: the categorical imperative itself cannot generate contingent ethical principles, it can only be used as a kind of test for principles which already exist. In this way, Kant's position is a critical one. He does not actually know what the right or the good is, but he knows which strategy of thinking to follow in order to attempt to produce it. The categorical imperative thus does not provide us with a substantive ethics, but it does urge us to adopt a certain attitude.

One can try to apply the same logic Kant uses to the problem of complexity and its ethical implications. The central characteristics of complex systems we have to consider are the following:

1. our knowledge of complex things is radically contingent in both time and space,
2. any decision we make concerning something complex has to be irreducibly provisional, yet
3. we have to act in a way which distinguishes the action from its alternatives otherwise we are not acting at all,
4. meaning emerges through the mutual interaction (both constraining and enabling) amongst components in the system, not through some pre-defined essence. Thus, as subjects we are constituted through interactions with others (both human and non-human) around us. My state depends on the state of others.

These principles can, following the Kantian logic, be reformulated in the form of an imperative. The following are possible ways of doing it:

1. justify your actions only in ways which do not preclude the possibility of revising that justification,
2. make only those choices which keep the possibility of choice open,
3. your actions should show a fundamental respect for difference, even as those actions reduce it,
4. act only in ways which will allow the constraining and enabling interactions between the components in the system to flourish.

These imperatives suffer from exactly the same shortcomings as Kant's categorical imperative. Nevertheless, they are not empty and can prepare the ground for a more detailed development of an ethics of complexity, an ethics which would be, like that of Kant, a critical one.

The Critical Position

The general importance of a critical position should also be developed in detail. It is important to show that such a position does not entail negativity or inaction, but that it is nevertheless critical to remain perpetually critical. This elaboration still has to take place, but three characteristics of such a position can serve as a platform for it.

A critical position informed by complexity will have to be transgressive. It can never simply re-enforce that which is current. Transformation takes place continually, despite all efforts to contain it. In this respect, I would argue, we need some bold alternatives to the orthodoxy of liberal democracy and free-market economy. Given the fact that communism failed, we should not allow ourselves to bullied into believing that the alternative is correct or much better. We should resist the macho nature of most political and economic cultures, irrespective of whether it is politically correct to do so or not. We should not be coerced, frightened or shamed into a state where we relinquish being transgressively critical.

A critical position will, in the most positive sense of the word, be an ironical position. There is no final truth which operationalises our actions in an objective way. Irony also implies, in a very systematic way, a self-critical position. Given the horrors of the world, this claim may be controversial. Nevertheless, we require a sense of humour if we are not to lose our humanity.

In the third place, a central role for the imagination is indispensable when we deal with complex things. Since we cannot calculate what will or should happen, we have to make a creative leap in order to imagine what things could be like. Aesthetic and creative activities are thus not interesting diversions, they open up the possibility of imagining better, more sustainable futures. Our humanity, our very existence, depends on this.

Literature

Byrne, D. (2005). "Complexity, configurations and cases." *Theory, Culture & Society,* 22(5): 95–111.

Cilliers, P. (1998). *Complexity and postmodernism: Understanding complex systems.* London: Routledge.

Cilliers, P. (2000). "Knowledge, complexity and understanding". *Emergence,* 2(4): 7–13.

Cilliers, P. (2001). 'Boundaries, hierarchies and networks in complex systems'. *International Journal of Innovation Management*, 5(2): 135–147.

Cilliers, P. (2005). "Complexity, deconstruction and relativism". *Theory Culture & Society*, 22(5): 255–267.

Cilliers, P. (2007). "On the importance of a certain slowness", in: C. Gershenson, D. Aerts and B. Edmonds (Eds), *Worldviews, science and us: Philosophy and complexity*. Singapore: World Scientific: 53–64.

Cilliers, P. and de Villiers, T. (2000). 'The Complex "I",' in *The Political subject,* edited by W. Wheeler, London: Lawrence & Wishart 226–245,

Cornell, D. (1992). *The Philosophy of the Limit.* London: Routledge.

Cornell, D., Rosenfeld, M. and Carlson, D.G. (eds.). (1992). *Deconstruction and the possibility of Justice.* London: Routledge.

Derrida, J. (1976). *Of grammatology.* Baltimore: The John Hopkins University Press.

Derrida, J. (1978). *Writing and difference.* Chicago: Chicago University Press.

Derrida, J. (1992). "The Force of Law: The "Mystical" Foundation of Authority", in Cornell, D., Rosenfeld, M. and Carlson, D.G. (eds.). (1992). *Deconstruction and the possibility of Justice.* London: Routledge.

Emmeche, C. (1997). "Aspects of complexity in life and science". *Philosophica,* 59(1): 47–68.

Held, D. (1980). *Introduction to critical theory: Horkheimer to Habermas.* Berkeley: University of California Press.

Kant, I. (ed. HJ Paton). (1948). *The Moral Law.* London: Hutchinson

Khalil, E. L. and Boulding, K. E. (Eds.) (1996). *Evolution, order and complexity.* London: Routledge).

Luhmann, N. (1989). *Ecological communication.* Chicago: University of Chicago Press.

Midgley, G., Munlo, I. and Brown, M. (1998). "The theory and practice of boundary critique: Developing housing services for older people." *Journal of the Operational Research Society,* 49: 467–478.

Morin, E. (2007). "Restricted complexity, general complexity", in: C. Gershenson, D. Aerts and B. Edmonds (Eds), *Worldviews, science' and us: Philosophy and complexity.* Singapore: World Scientific: 5–29.

Morin, E. (2008). *On Complexity.* New Jersey: Hampton Press.

McLennan, G. (2003). "Sociology's complexity". *Sociology,* 37(3): 547–564.

Novotny, H., Scott, P. and Gibbons, M. (Eds). (2001). *Re-thinking science.* Cambridge: Polity Press.

Pattee, H. H. (1973). *Hierarchy theory: The challenge of complex systems.* New York: George Braziller.

Simon, H. A. (1962). "The architecture of complexity: Hierarchic systems." *Proceedings of the American Philosophical Society,* 106(6): 467–482.

Toulmin, S. (1990). *Cosmopolis: The hidden agenda of modernity.* New York: Free Press.

Zellmer, A. J., Allen, T. F. H. and Kesseboehmer, K. (2006). "The nature of ecological complexity: A protocol for building the narrative", *Ecological Complexity,* 3: 171–182.

2

The Contributions of Social Constructionism to the Study of the Ethical, Social and Political Aspects of the Life Sciences

Erica Haimes

Introduction

This chapter addresses the question of what social constructionism (and its variations) can contribute to the study of the social, ethical and political aspects of genomics in particular and of the life sciences in general. I start by outlining the main features of constructionism in the social sciences and then discuss their usefulness and limitations. The latter mainly arise over constructionism as a distinct epistemological framework. I provide a brief example of work that has used the methodological resources provided by social constructionism to show, nevertheless, what might be gained from these approaches. I conclude by discussing the 'politics of constructionism',[1] exploring the relationship between the descriptive and the normative in this body of work and the extent to which they assist the possibility of an enhanced democracy in debates in genomics and the life sciences.

Social science and social constructionism: the roots of social constructionism

The epistemological roots of constructionism and the strength of its claims to knowledge can be debated at length. One commonly cited

---

[1] Velody and Williams, 1998

root is the seminal work by Berger and Luckmann, *The Social Construction of Reality* (1967) whose aim was 'not to produce a sceptical view of reality but rather to establish the grounds on which we claim certainty for our knowledge of reality'.[2] However in much of the later work on constructionism this subtle but important distinction has been lost and the sceptical view has dominated.

Broadly, constructionism has been popular since the 1980s when a proliferation of studies of the 'social construction of…' (variously race, religion, sexuality, identity etc.) emerged, leading Hacking to ask, almost in exasperation, 'The Social Constructivism of What?' (1999). Kenneth Gergen, the social psychologist influential in the development of constructionism saw it as 'not another foundational theory of knowledge but an anti-foundational dialogue'.[3] Generally constructionists are committed to relativism and ontological scepticism though beyond these two common elements there is 'a loose assemblage' of work.[4]

Delanty and Strydom position constructionism/ constructivism within a range of social science approaches that were part of a 'turn to context' in which the 'demarcation of science from society, the basis of positivism, becomes increasingly difficult'.[5] They argue that the debates that are central to these currents from the 1980s onwards are those between constructivism and realism and between relativism and universalism. They very nicely identify the following twelve elements as the general trends that inform and shape late twentieth century/early twenty first century social science epistemology and methodology. Their relevance to an exploration of the issues for democracy and genomics becomes clear from even a brief glance:

1. knowledge is historically embedded.
2. truth is relative: "the truths of science can no longer be seen as timeless, transhistorical and universal but relative to the scientific context and to the historical location of science".[6]

---

[2] Velody and Williams, 1998, p. 4
[3] 1995, p. 20, cited in Velody and Williams, 1998, p. 2
[4] Ibid, p. 3
[5] Delanty and Strydom, 2003, p. 365
[6] Delanty and Strydom, 2003, p. 366

3. the decline of the neutrality of science, to which they add, "It is increasingly recognized…that public debate has to be allowed to play a much greater role in science today".[7]
4. the constitutive nature of theory: the view that science constructs the reality that it purports to describe.
5. knowledge is socially contextualized: "the turn to context – under the rubric of locationalism, situationalism, contextualism – is one of the main developments in the epistemology of social science in recent decades".[8] Delanty and Strydom point out though that debates continue as to how far contextualism should be taken: "In the more extreme claim, as in certain kinds of constructivism, it is held that all of science is constructed by social actors and that there is no difference between science and other forms of knowledge".[9]
6. epistemological uncertainty: "the historical, linguistic and cultural, and more broadly the reflexive, turn in the philosophy of the social sciences has introduced an unavoidable degree of epistemological indeterminacy and uncertainty into the core of science…scientific knowledge suffers the same fate that has fallen on most, if not all, normative discourses…and as Jurgen Habermas insists, the experts have been drawn into legitimation controversies in which publicly acceptable reasons prevail rather than authoritative knowledge".[10]
7. the end of physicalism: that is, "the end of there being a fixed order of reality external to science".[11] In constructionist literature elsewhere, we find debates about the 'realist residue', that is, how far does constructivism go in asserting the nature of the material and in deciding what, if anything, is 'unconstructed'. Haimes and Williams (1998) suggest these questions are best examined within the context in which they are debated since the terms in which they are debated (e.g.

---

[7] Ibid, p. 366
[8] Ibid, p. 367
[9] Ibid, p. 367
[10] Ibid, pp. 367-8
[11] Ibid, p. 368

the distinction between the 'natural' and the 'social') and the definition of those terms, tend to be context-specific. The distinction cannot be asserted other than through empirical examination of the historical context in which they occur. They illustrate this through an analysis of the UK Human Fertilisation and Embryology Authority's (1994) public consultation document on the use of fetal tissue and female cadavers as sources of ovarian tissue to provide eggs for treatment and for research.

8. Anti-reductionism: "the recognition of the manifold nature of reality or society and the multiplicity of possible interpretations".[12]
9. Contingency: "science, as just one activity among others, is surrounded by critical discourses which relativize it and bring the public, in an increasingly significant way, into play".[13]
10. Anti-essentialism: "commits social science to the view that social reality is processual, in flux and multi-faceted".[14]
11. The world is best seen as an artefact
12. The decline of disciplinarity: 'research [is] driven more by problems, ethical-political concerns and policy-directed programmes than by disciplinarily specific traditions'.[15] This view reflects much of the work done in the UK on ELSI and genomics where multidisciplinary approaches have emerged and are increasingly advocated for gaining effective analytical legitimacy,[16] much as the ethnographer gains legitimacy by 'immersion in the field'.[17]

A number of social science approaches come under this description, many of which have their roots in older traditions of interpretative analysis (drawing on the work of writers such as Schutz, Simmel, Weber, Geertz, Cicourel, Garfinkel, Goffman) and developing into schools such as ethnomethodology, grounded theory, cognitivist

---

[12] Delanty and Strydom, 2003, p. 368
[13] Ibid, p. 368
[14] Ibid, p. 369
[15] Ibid, p. 369
[16] Haimes, 2006
[17] Wellin and Fine, 2001

analysis, symbolic interactionism, phenomenology, from which 'social constructionism' emerged.

Types of Constructionism

Following on from this excellent summary of the possible theoretical, epistemological and social positionings of constructionism, Delanty and Strydom then follow a fairly conventional line by identifying three kinds of constructionism: social constructionism, social constructivism (or scientific constructivism), and radical constructivism.[18] This typology smoothes over the variations asserted by Velody and Williams (1998). Delanty and Strydom identify their first type, social constructionism, as developing out of the Berger and Luckmann text and the view of its proponents (identified as, among others, Cicourel, Garfinkel, Goffman and to some extent Hacking) who "simply assumed' that social reality is 'continuously being in a process of creation by human agency".[19] There is of course nothing 'simple' about such an assumption.

Delanty and Strydom regard 'social constructivism' as a stronger version of social constructionism as it asserts the "central claim that science is constructed by social actors…not merely influenced by social factors" and a view of there being 'no essential difference between nature and society'.[20] An influential body of work has arisen from this central thesis, variously labelled as Science and Technology Studies (STS) and Sociology of Scientific Knowledge (SSK), with key authors such as David Bloor, Karin Knorr-Cetina, Bruno Latour, Steve Woolgar, Sheila Jasanoff. 'Radical (cognitivist/operative) constructivism' refers to a very specific strand of work developed mostly by Niklas Luhmann drawing together cognitive biology and cybernetics to focus "on the processes of construction in which a system engages in relation to its environment".[21]

---

[18] Ibid, pp. 373-4

[19] Delanty and Strydom, 2003, p. 373

[20] Ibid, p. 373

[21] Ibid, p. 374

Such characterisations, though useful, can be challenged. For example, in discussing the very particular issue of how we might think of the relationship between the biological and the social Haimes and Williams,[22] suggest the following way of characterising the relevant traditions: (a) historians of ideas, such as Ludmilla Jordanova, who regards the nature/culture distinction as a cultural artefact, whereby she writes of the 'languages of nature', showing how ideas around biology have been entangled in, and co-implicated with, ideas of medicine, science, art and literature. Nonetheless such approaches tend to deal with 'representations', rather than constructions, of nature; (b) contributors to the Sage *Inquiries in Social Construction* series such as Gergen and Shotter who take a more idealist and relativist approach and argue that "the language of social description cannot be mapped onto stable external realities: language is the only reality we can know"[23]; (c) a similar group to Delanty and Strydom's social constructivists, who discuss the social construction of science. For these, the facts of nature, biology and of any science are given the status of cultural objects.[24] Latour and Woolgar assert, "We do not conceive of scientists using various strategies as pulling back the curtain on pre-given, but hitherto concealed truths. Rather objects…are constituted through the artful creativity of scientists".[25]

Michael Lynch provides a "constructivist genealogy of social constructivism" and concludes that the use of the word 'construction' provides a

> productive confusion…It is productive because it provides an initial hook that captivates adherents, encouraging them to invest their theoretical, methodological and political hopes in an academic approach or movement. But it can be confusing when adherents try to replace the eclectic surface affinity which fostered the success of constructivism in the first place, with something deeper and more coherent.[26]

---

[22] Haimes and Williams, 1998, pp. 136-7

[23] Ibid, p. 137

[24] Knorr-Cetina, 1993

[25] 1979, p. 128, cited in Haimes and Williams, 1998, p. 137

[26] Lynch, 1998, p. 29

Of the three elements that he highlights as the attractions of constructionism (theoretical, methodological and political) it is the theoretical that is most open to question, whilst the methodological and the political have most to contribute.

Science and technology studies

Whilst there are some problems with the social constructivist / SSK approach it is nonetheless the constructionist approach that is now most commonly referred to in discussions of the life sciences. Hess (2001) outlines the methodological approach of science studies and provides a subtle delineation, particularly between first and second generation, within this school of work. Hess' view is that the first generation of SSK work arose in the 1980s and is the equivalent of Delanty and Strydom's social constructivist 'strong progamme'. This work was (broadly) populated by researchers such as Collins and Pinch, Latour and Woolgar, Callon, from whom ideas such as co-construction and actor network theory emerged. Of interest in this work is the idea of how non-human entities gain a 'delegated agency' within sociotechnical networks through which human action is shaped.[27] However some of this work, Hess suggests, was associated with excessive claims about the nature of the material world which in turn led to strong reactions amongst philosophers and scientists (and to the ensuing 'science wars').

Hess then describes a second generation of STS work which he regards as having "a more American flavour" and as being more diverse including anthropologists, feminists, and cultural studies researchers. He suggests this work is more oriented towards social problems and more likely to study settings outside the laboratory, to study "lay groups, activists, social movements, the media and popular culture" and to be more interested in the emergence of orthodoxies and heterodoxies in a field, to examine variations in expert knowledge, and to work across cultures.[28] However, his characterisation of the second generation is so broad that I am not sure that all that he includes would be seen by the writers named as

---

[27] Hess, 2001, p. 235

[28] Ibid, p. 236

necessarily STS work and might more appropriately be considered as good ethnography within settings of science and medicine. Hess sums up the differences between the two generations:

> It is probably more accurate to say that in the second generation the construction problems shifts from the SSK focus on how social and technical factors are interwoven in knowledge and technology production (social construction) or how sociotechnical networks and societies are mutually constituted (co-construction) to how cultural meanings or legitimating power relations are embedded in science and technology (cultural and political construction) and how different actors interpret science and technology (reconstruction).'[29].

Hess suggests that the difference between the two generations is that the second generation take a 'both-and' view of the co-determination of nature and culture rather than an 'either-or' approach.

An increasingly influential strand of work that in many ways straddles Hess' generations is that of 'co-production' most closely associated with the STS work of Jasanoff and colleagues (2004). The central thesis of this work is that "we gain explanatory power by thinking of natural and social orders as being produced together...Briefly stated, co-production is shorthand for the proposition that the ways in which we know and represent the world (both nature and society) are inseparable from the ways we choose to live in it'.[30] Common preoccupations of this work include how new developments are recognised and established, the terms on which disputes are based and considered resolved, the cross-context character of science and technology, and the socio-cultural practices and institutions of science and technology.[31] (Echoing a point already made, Jasanoff argues that co-production 'fits most comfortably with the interpretive turn in the social sciences, emphasizing dimensions of meaning, discourse and textuality'.[32]However, in making claims about the links and possibilities that co-production as an explanatory (note, not

[29] Ibid, p. 237
[30] Jasanoff, 2004, p. 2
[31] Ibid. p. 5-6
[32] Ibid. p. 4

descriptive) resource provides for other disciplines one is left with the question of what is not included in, or answered by, co-production?

The cultural relativism of the second generation approach makes ethnographic research a particularly appropriate methodology. However, for Hess, ethnographies in second generation science and technology studies face unique problems that traditional anthropological ethnographers tend not to have to confront. First, the STS ethnographer is studying a field that is a part of his/her own society; second, knowledge in that field can often be new and changing leaving both ethnographer and informants unsure of the status of that knowledge; third, the STS ethnographer is working with people who are very likely to read his/her analyses and can feel threatened by the very fact of the study.[33] Nonetheless the STS ethnographer has to achieve the same standards of 'good' ethnography as any other ethnographer: they have to achieve native competence in the technical aspects of the science and technology involved; be able to interrogate the taken-for-granted features of that world, and make a new empirical or theoretical contribution to social science.[34] He adds another criterion that emerges from the social problems approach of the second generation of studies: that is, to be active as 'citizen-researchers', to facilitate dialogue between the field of research and the wider public. Hess suggests that the second generation STS ethnographer is a 'post-constructivist', interested not only in how knowledge and technology are constructed but also how they might be better constructed, providing alternatives and actively intervening in policy debates.[35] We can see that such a suggestion is taking social science through a 'normative turn', a point returned to in the Discussion.

Criticisms of constructionism

There is much to commend itself in Hess' second generation of science and technology studies particularly in methodological terms but it is worth making a brief visit to the criticisms that have been

[33] Wienroth, 2008
[34] Hess, 2001, p. 239
[35] Ibid, pp. 239-242

voiced about constructionism as an overall approach. The strongest critique comes from realist schools of thought. Delanty and Strydom see the debate between realism (whose proponents include Rom Harre, Roy Bhaskar, Ted Benton, John Urry) and constructivism as mirroring the older debate between understanding and explanation in the social sciences, with constructivism placing the emphasis on understanding through interpretation and realism placing the emphasis on explanation. The notion of realism is 'based on the assumption that an external reality exists which is independent of human consciousness yet can nevertheless be known' without however assuming that it is easily observable and explained through causal analysis.[36] Velody and Williams acknowledge that writers like Roth and Turner criticise constructionists for their relativist pretensions,[37] whilst pointing to the often non-relativist style of their conclusions (indeed we could perhaps say 'explanations' here).

There are however elements of common interest between realism and constructionism, as we can see when we consider the breadth of work that is included under the constructionist banner by writers such as Hess. The notion of an emergent, layered reality in realism,[38] is not that different from the 'emergent definitions' of ethnographic inquiry or the 'constant flux' of ethnomethodological versions of constructionism. As Delanty and Strydom suggest, constructionists 'are not saying that reality is imagined or does not exist, but that reality exists in science only in ways defined by science.'[39] Here they do not distinguish between constructivists but one might suggest that this refers to constructivists other than the strong programme of SSK work. They also suggest that Hacking's work indicates that the line between realism and constructivism is also 'very thin'.[40]

Hacking's work is interesting since his criticisms of constructionism come from someone who has a great deal of sympathy with this perspective. Initially he appears to be damning: "The trouble is that, except as a political slogan, the very phrase "social constructionism" is tired, imprecise, pretentious and self-

---

[36] Delanty and Strydom, 2003, p. 376
[37] Velody and Williams, 1998, p. 3
[38] Delanty and Strydom, 2003, p. 376
[39] Delanty and Strydom, 2003, p. 377
[40] Ibid, p. 377

righteous".[41] Like Lynch, Hacking views constructionism as a diverse assortment of approaches, "a kettle of many very different kinds of fish".[42] His frustration about social constructionism is its very imprecision: it has become a slogan, interested in unmasking but without being clear about what it is that is being unmasked, given the tendency in such work to use broad terms to cover a wide field: the social construction of science, gender, race, age etc. In some ways Hacking is an example of Hess' second generation of social constructionists as he addresses a number of fields (e.g. child abuse, teenage pregnancy) outside science studies. However much of what he says, both in favour of constructionism and in criticism of it, applies as equally to science as to other fields. He pushes for analysts to be more precise in their inquiries.

> It is not prudent…[to say] that child abuse is both real and constructed. Instead say that the behaviour we now call child abuse has been a fact of most human communities at most times in our history. Say that the idea of child abuse, and the consequences of that idea, were constructed in the course of the years following 1960. I think that "making and moulding" is a better term than "constructing". Often it is better to be still more specific'.[43]

The elements of constructionism that Hacking favours are those that show that particular ways of dividing up categories of persons or behaviours are not inevitable; that we consider these persons or behaviours as 'worthy' of their own labels 'is a fact of a thoroughly contingent social history'.[44] Part of this process is the unmasking that is often a part of constructionist analysis: showing how concepts and labels are generated and impose or encourage a certain order or way of thinking on a field of practice. Hacking wants to draw attention to the historical circumstances in which certain 'practices, institutions and experiences' come into being and which, in turn, facilitate the creation of certain labels that 'identify, advise, control' certain social

---

[41] Ibid, p. 426
[42] Ibid, p. 421
[43] Ibid, p. 423, emphasis in original
[44] Ibid, p. 426

groups with often serious consequences – but says 'there are better ways of saying that than by talking about social construction'.[45]

Beyond constructionism or a return to roots?

I share some of Hacking's sense of fatigue with the uses of the term 'constructionism', whilst sharing the view that detailed studies of how certain practices, ideas and institutions arise is very important. Equally I share with Hess the view that the ethnographic approaches of science studies have much to offer, through their detailed examinations of the settings of science and medicine. What he describes as second generation science studies show how much insight can be gained from trying to understand aspects of science and medicine from the detailed perspectives of those directly involved in those settings as practitioners, patients, research subjects and so on. Therefore it is not so much constructionism as a slogan or science studies as a label that provide insight but rather the methodological approaches that both adopt as a resource for doing their work that shows us what can be gained. These methodologies derive from the interpretavist underpinnings that both share (also referred to elsewhere as 'ethnomethodological ethnography'[46]) within the broad epistemological locations identified by Delanty and Strydom.

I am in sympathy with Lynch's view of the 'origins' or genealogy of social constructionism. He argues that the main influence of Berger and Luckmann's analysis of the 'social construction of reality' is simply the phrase 'social construction'. Otherwise their account was less radical, he argues, than many of their contemporaries in the late 1960s who did not use the phrase but who have nevertheless proved to be more influential on work that is now referred to as constructivist. He refers to 'the diverse, overlapping array of semiotics, ethnomethodological, symbolic interactionist, radical feminist, cultural-Marxist, literary theoretic and discourse analytic

[45] Ibid, p. 427
[46] Silverman, 1985

approaches'.[47] He also credits Berger and Luckmann with developing a version of Schutz's social phenomenology that integrated with more mainstream social theory (see Delanty and Strydom's list of features). Schutz, Lynch argues, was interested in explicating 'the natural attitude of an ideal-typical member of a society or group…not in developing an explanation or causal theory'. His work resonated with others such as Garfinkel and his ethnomethodological approach and built upon Weberian interpretative sociology and notions of the importance of subjective understanding as a 'constitutive property in the social world that had profound implications for any description of it'.[48]

Whilst some critics, particularly contemporary STS scholars, see this as abandoning a view of 'society-as-a-whole' in favour of micro analyses (and therefore seen by some as 'necessarily' trivial), Lynch points out 'this de-transcendence was designed to bring into relief how anybody, whether a prominent social theorist or nobody in particular, manages to take account of an intersubjective order while acting constitutively within it'.[49] This move away from the 'craving for generality',[50] towards what I would call 'analytical descriptions' means that 'a whole array of practical, locally organised, linguistically expressed ways of making ordinary judgements about matters of fact, truth, credibility and reality become available' for examination.[51] Perhaps some STS scholars fear that if they do not move from empirical data to make 'grand statements' about society-as-a-whole, they and their research will not be taken seriously.

A recent example

To understand what these debates amount to in practice, it is useful to view a brief example of constructionist work. For Haimes and Williams, 'constructionism is neither an epistemology nor a sociological substitute for epistemology but a technique or mode of

---

[47] Lynch, 1998, p. 24
[48] Ibid, p. 25
[49] Ibid, p. 26
[50] Wittgenstein, 1958, p. 18, cited in Lynch, 1998, p. 26
[51] Ibid, p. 26

analysis'.[52] The question then is how does the researcher best 'get into' or understand what is going on in this setting or this context, or perhaps more accurately, how does the researcher get a plausible version of what might be going on? And how does the researcher warrant his/her claims to be providing a plausible account whilst acknowledging that other accounts are possible (as long as they too are warrantable). In order to achieve this, the researcher has to focus on the particular rather than the general: one has to be specific as Hacking argues.

In a recent special issue of the journal *New Genetics and Society,* looking at the social worlds of stem cell science, several pieces of constructionist-style research explore the practices around the provision of reproductive tissue for human embryonic stem cell (hESC) science. In one article the authors describe how their two related UK and Swiss projects, studying embryo donation for human embryonic stem cell (hESC) research from IVF patients, indicate the variability of meanings that actual and potential donors ascribe to their embryos as they go through, and then reflect on, their IVF treatment.[53] One simple but crucial distinction that potential donors draw is that between '"the embryo" as an abstraction with an emblematic moral status; "an embryo" which could be any particular embryo that results from the IVF process, and "our embryo"…imbued with particular social and moral values by the couple being treated because "our embryo" might become "our baby"'.[54] Such variability in meaning is also evident in Swiss law, showing that it is not just patients or lay persons who work with multiple meanings. What is also clear is that the embryo has to be posited as a particular sort of entity, by particular groups of people, in particular contexts, in order for them to be used in particular ways (as 'precious' entities with the capacity to produce a baby; as a precious resource for research; as an item that is 'not-waste' and therefore useable for research, and so on). The two studies 'reveal that embryos are not fixed, universal biological entities but are defined by, and acted upon, in relation to their social context, that is, by their location in time and space…it is not possible to have an effective debate

[52] Haimes and Williams, 1998, p. 4
[53] Haimes et al, 2008
[54] Ibid, p. 124

about the acceptability of embryo research if discussion is restricted to the moral status of the abstract embryo; the real physical entities that reside in the freezer, clinic and women's bodies need to be considered',[55] as do the views of the embryo donors themselves.

In the same issue, Williams et al draw on two ethnographic studies of the social practices of pre-implantation genetic diagnosis and of hESC science to understand 'the social construction of embryos'[56]. They use the ethnographies to unpack the variations in meanings of the human embryo in these different contexts and conclude that the embryo operates as a boundary object. They argue that this study provides an example of 'the production of biomedicine as a social and material practice in the emerging sphere of new medical technologies'.[57]

What these studies show (along with the study mentioned earlier of the HFEA consultation document) is that 'the biological' and 'the social' are boundary marking and boundary permeating concepts and the division between them is shifting and unstable. Therefore it is necessary to 're-specify' the distinction as one which is capable of empirical examination in particular historical contexts, rather than take a prior stance on the ontological character of the natural and the social. Therefore, the strong programme in science studies, that makes 'radical claims for the autonomy of the social and its appropriation of the biological',[58] can be seen to create some difficulty, both for social theorists and for everyday actors in the practical world.

Discussion: The politics of 'constructionist' work

None of the above studies of the meanings of embryos could be dismissed as dealing only with micro level trivia. Nonetheless that is a concern often expressed about such detailed contextualised studies: that they cannot lift their analyses above the context, to consider

---

[55] Ibid, p. 124
[56] Ibid, p. 7
[57] Ibid, p. 7
[58] Haimes and Williams, 1998, p. 137

society-as-a-whole. However, as has already been indicated by Hess and, in different ways, by Jasanoff, for many researchers the political power of constructionist-style work is one of its main attractions.

Velody and Williams (1998) refer to two aspects of impact of constructionism on politics: first, the politically informed nature of the growth of constructionism (to which we can add Hess' second generation work and Jasanoff's 'States of Knowledge'); second, the ways in which constructionist analysis unmasks the practice of politics itself. For example, we can see, from an analysis of the 2008 UK parliamentary debates, the processes of narrative construction of the 'woman egg sharer' for somatic cell nuclear transfer research and the deliberate entangling of this social category with that of the 'hybrid embryo' as a strategy to persuade fellow parliamentarians of the necessity for hybrid embryos to 'save' the egg sharer from distress.[59] More generally constructionist analysis shows variations in meanings and diversity of sense making practices, that consequently destabilise views of what can be taken for granted (or as I prefer it, taken as asserted) or as self-evident.

A third aspect should be added: the ways in which constructionist work often involves adding to the voices that are most often heard in these debates, by expanding our awareness of the views and opinions of those beyond the usual stakeholders. In genomics and the life sciences the key stakeholders are often identified as scientists, governments, corporations, non-governmental organisations, and religious or secular faiths. As we have seen above, a constructionist exploration of the world of genomics expands that list to include clinicians, research funding bodies and, perhaps most important of all, non-professionals (or 'lay' groups, though both are unsatisfactory terms[60]). This last group can be sub-divided: 'the directly involved non-professionals' (for example, patients, tissue donors, research participants, expected beneficiaries) and the 'non-directly involved non-professionals', such as the wider communities who might not have direct experience of these developments but who are nonetheless affected by, and have an effect on, how these

[59] Dowdle et al, 2009

[60] Rabeharisoa and Callon, 2004

developments proceed, in numerous expected and unexpected ways. These additions endorse the argument that there is a need to take a de-centred view of politics and to study the complex and dynamic networks of different social actors.

This last point suggests that it is important to look for and examine the interconnections between different stakeholders, and to look for the discontinuities and fragmentations within any particular stakeholder group. As social constructionism as outlined above suggests, this is best done within particular social contexts to see how these alliances and differences emerge, interplay, and are then perhaps reshaped within particular circumstances. In that way ideas about fruitfulness of friction between multiple stakeholders can be explored, as can the consequences of the absence of friction in some cases.

We can therefore divide the political impact of constructionism in two ways: intentional impact and 'by default' impact. The work of Flyvbjerg (2001) is very interesting in urging social scientists to develop work that is theoretically and methodologically robust, whilst being politically relevant. He advocates the use of a phronetic social science that shares many features with constructionist work whilst also helping to bridge some of the divide between social science and ethics. His recipe for such work includes: focusing on values; placing power at the core of analysis; getting close to reality; emphasizing little things; looking at practice before discourse; studying cases and contexts; asking 'how' and producing narrative; joining agency and structure, and encouraging dialogue amongst a polyphony of voices[61]. We have used Flyvbjerg's work to consider how an analysis of participation in genetic databases can be strengthened by taking seriously the sometimes conflicting approaches of social science and ethics, but what is of particular interest here, is that his approach aims to 'explicate precisely how individual and collective deliberations and actions are informed by value considerations as part of the lived texture of practical reasoning. Furthermore, to make knowledge of this texture available to others so that future deliberations are informed less by stipulations of what should matter to those who want to make wise judgements and more by understandings of the multiplicity and distinctiveness of how individuals are able

[61] Flyvbjerg, 2001, p. 140

accountably to determine right or wrong courses of action in specific social contexts'.[62]

It is through such strategies and those of the many other constructionists identified throughout this chapter that interpretavist-constructionist work can systematically tackle the big questions such as 'what is the nature of public trust and of democracy in the life sciences, and how can it be improved?'

Acknowledgements

I am grateful for the opportunity to participate in the International Expert Seminar on 'Towards a "lingua democratica" for the public debate on genomics' at the University for Humanistics in Utrecht in October 2008 for which the ideas for this chapter were first developed. I am grateful too for the encouragement and comments given by Peter Derkx, Harry Kunneman, Paul Cilliers, Cees Pieters and the other participants in the seminar.

Literature

Berger, P. and Luckmann, T. (1967). *The Social Construction of Reality*. London: Allen Lane

Delanty, G. and Strydom, P. eds. (2003). *Philosophies of Social Science*. Maidenhead: Open University Press

Dowdle, S., Haimes, E. and Taylor, K. (2009). "Egg sharers and hybrid embryos: gender, science and the process of category construction in UK parliamentary debates". Unpublished manuscript, Newcastle University

Flyvbjerg, B. (2001). *Making Social Science Matter*. Cambridge: Cambridge University Press

Hacking I. (1999). *The Social Constructivism of What?*. Cambridge, MA: Harvard University Press

Hacking, I. (2003). "What is social construction?", in Delanty and Strydom, eds *Philosophies of Social Science*. Maidenhead: Open University Press: 421-427

[62] Haimes and Williams, 2007, p. 466

Haimes, E. (2002). "What can the social sciences contribute to the study of ethics?". *Bioethics* 16: 89-113

Haimes, E (2006). "Cross-national insights: the UK", Presentation to the ERA-SAGE 'Comparing ELSA' Conference, Basel, Switzerland, November

Haimes, E., Porz, R., Scully, J. and Rehmann-Sutter, C. (2008). "'So, what is an embryo?" A comparative study of the views of those asked to donate embryos for hESC research in the UK and Switzerland'. *New Genetics and Society,* 27-2: 113-126

Haimes, E. and Taylor, K. (2008). "Fresh or frozen? Socio-ethical issues in acquiring and providing human embryos for embryonic stem cell research." Presentation to the International Association of Bioethics World Congress, Rijeka, Croatia, September

Haimes, E. and Williams, R. (1998). "Social constructionism and the new technologies of reproduction", in: Velody, I. and Williams, R. eds. *The Politics of Constructionism.* London: Sage: 132-146

Haimes, E. and Williams, R. (2007). "Sociology, ethics and the priority of the particular: learning from a case study of genetic deliberations". *British Journal of Sociology*, 58 (3): 457-476

Hess, D. (2001). "Ethnography and the development of science and technology studies", in Atkinson, P., Coffey, A., Delamont, S., Lofland, J. and Lofland, L. eds. *Handbook of Ethnography*, London: Sage, pp. 234-245

Human Fertilisation and Embryology Authority (1994). "Donated ovarian tissue in embryo research and assisted conception: public consultation document", London: HFEA

Jasanoff, S. ed. (2004). *States of Knowledge: the co-production of science and social order.* London: Routledge

Knorr-Cetina, K. (1993). "Strong constructionism – from a sociologist's point of view". Social Studies in Science, 23: 555-566

Latour, B. and Woolgar, S. (1979*). Laboratory Life.* London: Sage

Lynch, M. (1998). "Towards a constructivist genealogy of social constructivism", in: Velody, I. and Williams, R. eds. *The Politics of Constructionism,* London: Sage: 3-32

McLeod, C. and Baylis, F. (2007). "Donating fresh versus frozen embryos to stem cell research: in whose interests?". *Bioethics,* 21-9: 465-477

Rabeharisoa, V. and Callon, M. (2004). "Patients and scientists in French muscular dystrophy research", in: Jasanoff, S. ed, *States of Knowledge,* London: Routledge:142-160

Silverman, D. (1985). *Qualitative Methodology and Sociology.* Aldershot: Gower

Velody, I. and Williams, R., eds. (1998). *The Politics of Constructionism,* London: Sage

Wellin, C. and Fine, G. (2001). "Ethnography as work", in Atkinson, P. et al, eds. *Handbook of Ethnography,* London: Sage: 323-338

Wienroth, M (2008) *The Lowest Common Denominator?* Unpublished PhD thesis, Newcastle University

Williams, C., Wainwright, S., Ehrich, K. and Michael, M. (2008). "Human embryos as boundary objects?", *New Genetics and Society,* 27-1: 7-18

3

Engineering, Humanistics and the Lingua Democratica: On Cross-Disciplinary Communications between 'Strange Bedfellows'

Cees Pieters

Introduction

The NWO-programme on the "Societal Aspects of the Genomics Debate" (MCG) (NWO 2001) spawned various initiatives across the Netherlands in order to understand the impact of a new technology in society. This technology is related to engineering (human) genes, which is expected to have major consequences, both locally as well as on a global scale. Not only does genomics promise great advances in human health and well-being, its potential hazards are equally massive. Besides this, genomics touches the foundations of our biological existence, which raises questions and concerns of a deep philosophical and ethical nature. It is progress along a steep gradient, to use a very technical term.

The contribution that was supervised by the University for Humanistics, "Towards a Lingua Democratica of the Genomics Debate"[1] has focused on important stakeholders in this debate – governments, NGOs, research laboratories, companies and so on - and aimed to progress beyond the common uproars of mutual 'criticism of the others', and look for means of opening new, 'fruitful' interactions between the various domains. Within this program, I was given the opportunity to represent the hard sciences, being an engineer with a training and a long professional background in

[1] Kunneman, 2003

electro-technical and software engineering. This has made me particularly 'sensitive' to the problematic interactions between technologists and the humanities, both within the context of genomics as beyond this.

These interactions are rich in friction. From one end, there is a certain view on technologists which can be exemplified in the following excerpt taken from one of the final round-table conversations that was organised by the MCG in 2007:

> Hard scientists tend to invite ethicists into their labs and then think that they can present a beautiful image of science to the world. It is difficult for an ethicist to explain why you are not in favour of or against science, but that you are interested in the underlying processes. [...] it is of importance that the humanities and artists are experienced in their social interactions with scientists [...]. They have to be very conscious of the dangers and resist them.[2]

Apparently the intentions of ethicists cannot be conveyed easily. They need to be contested and defended, for there are 'dangers' that need to be 'resisted'.

Conversely, the technological side can be equally problematic. In his admirable coverage of (philosophical) 'thinking through technology', Carl Mitcham highlights some stances from technologists:

> Technological philosophy can criticize humanities philosophy of technology for being too speculative or based on too narrow if not unempirical foundations. Humanities thinkers do not understand what they are talking about, say engineer-philosophers.[3]......C.P. Snow, for example, once contrasted the scientific and literary intellectuals, arguing that the former know more about Shakespeare than the latter know about thermodynamics. Engineering curricula require engineering students to take humanities courses, but how many engineering courses are humanities students required to take?[4]

---

[2] Beintema and Vrouwe 2007, p. 7, translated from Dutch by author

[3] Mitcham 1994, p. 64

[4] Ibid, p. 64

It is friction that surfaces more often when reading literature on technology and philosophy or ethics. These interactions proceed with a wary politeness at best, and an interesting blend between mild contemptuousness and disinterest at worst[5]. Engineers and philosophers, as Goldberg argues, make "strange bedfellows"[6].

From the perspective of the 'lingua democratica for the genomics debate', this divide deserves serious attention. Technology is a magnificent force of progress that –and both the enthusiasts and the critics of technology agree on this- seems to have an autonomous drive that seems to go faster and faster[7], but it is not at all sure whether this 'progress' is what we –as human beings- really want or need, especially since it often seems to go hand in hand with unforeseen new problems or implications. To borrow a fitting term from the American TV show "Dexter", technological progress seems to travel with a "dark passenger".[8] Although this stowaway often takes the ride in silence, its tantrums can shake the world, and even challenge our existence.

I have been looking into this friction from a 'bricolage' of scientific disciplines that is currently more and more called 'complexity' or 'complexity thinking'.[9] In this contribution I therefore aim to explore this chasm between technologists and humanities thinkers from this perspective, by looking into some of the systemic issues that are related to the production of theory.

Technology and Philosophy

When looking into a 'lingua democratica' between two stakeholders, one needs to keep in mind that this, logically speaking, introduces four angles of reflection. Two of them usually get sufficient focus, and follow a path from 'self' to the 'other', usually with a focus on the latter. On issues relating to technology, philosophical and ethical reflection on technology and technologists are –apparently after a

---

[5] Pieters, 2007
[6] Goldberg, 2007
[7] Schmidt and Marratto, 2008; Kurzweil, 2006
[8] Siega, 2007
[9] Waldrop, 1992; Cilliers, 1998

very slow start- now sufficiently being covered.[10] Conversely, although perhaps not very obvious, engineering perspectives on philosophy (through technology) also have a respectable heritage.

The two remaining perspectives are related to reflection on 'self'. These perspectives shape and maintain the identities of the stakeholders in their attitude towards technology, their perceived roles in society and the contributions they purport to offer.

If one assumes that an human agent, or groups of human agents, host(s) domains of inherently limited and constrained knowledge, then all these perspectives can considered to be shaped by certain bias.[11] This is necessary, for it assists the observer in differentiating 'essence from details'. This necessary, purposeful reduction helps to shape observations and argumentation. However, they also skew and distort the perceptions of our life-world by accentuating certain issues, under guidance of implicit and explicit interests and values.

In general, one can say that reflection of is not a high priority of engineering practice. Engineering tends to focus on process, on changing the world from one that has certain perceived problems, to a 'better' world without them. Humanities reflection on technology, on the other hand, often is just that –reflexive-, and usually includes careful scrutiny on the technologists. This often results in a 'one-way' perspective that aligns with their criticism, and to which technologists hardly respond if they are at all aware of the critiques. The resulting 'Amsterdam-syndrome'[12] implies that there is no real communication, and therefore there is no lingua democratica to speak of.

One of the most striking differences between humanities philosophy of technology and engineering can be seen in the way theory is used in either field. Engineering is more or less a praxis science, in which the experimental method that was perfected by Newton plays a dominant role.[13] This method introduces a fundamentally different way of dealing with theory, with consequent systemic differences, in comparison to predominantly theoretical contributions. This issue

---

[10] Ihde, 1993; Mitcham, 1994

[11] Rassin, 2007

[12] Pieters, 2007

[13] White 1999, p. 182

becomes very important when the 'common ground' to engage in dialogue is that of complexity, because here theory can be considered to be the result of a production system that generates new theories from existing ones. In other words, the art of theorising can inherently be considered a self-describing, and thus complex, production system.

In order to understand the chasm between theorists and technologists, it may therefore be good to take a closer look at these systemic differences behind the production of their intellectual brainchildren. I will concentrate here on one issue that I think contributes to the gap at a very fundamental level, and goes back to the differences between theoria and techne.

Theoria and Techne

If all produce of human thinking is biased, then is will not be a surprise that the American philosopher of technology Don Ihde observes certain theory-bias in many narratives of science and philosophy regarding technology:

> That bias points to much deeper and longer set of prejudices which lay behind the late arrival of our subject matter, the philosophy of technology.[14]

Ihde argues that this prejudice goes back to the notion of theoria of the ancient Greeks:

> But, at the heart of theoria in the Greek sense was also a kind of quasi-religious belief which associated the 'highest' virtues with a contemplative attitude.[15]

Theory-bias has long obscured the important role that techne played in the evolution of 'natural philosophy' into contemporary science. Only in the course of the Seventies of the previous century it became more and more clear that theoria and techne had bonded into a "symbiotic relationship"[16] that Bruno Latour has called technoscience. This symbiotic relationship was kick-started by the

---

[14] Ihde 1993, p. 19
[15] Ibid, p. 20
[16] Winston and Edelbach, 2008, p. 12

experimental method of, amongst many others, Galilei, Keppler and especially Isaac Newton.[17]

In this view, it was the synthesis and consequent co-evolution of instrumentation and (especially formal and mathematical) theory that became the 'engine' that drives technoscience, as instrumentation opened new ways of observing and understanding the world. The invention of the microscope introduced new intellectual concepts such as 'bacteria', 'viruses' and eventually DNA; conversely the telescope brought us 'galaxies' and 'red dwarfs'. These novel ways of observation spurred the development of theory which, in turn, resulted in new instruments. The resulting positive feedback loophas been driving the generator that has powered the Industrial Revolution, mass-production and the information age. [18]

Morton Winston argues that social values have also been a partner in this symbiotic relationship of technoscience. Early Enlightenment saw a shift in thinking of the relationship between humans and nature from one of submission to that of domination, as is reflected in the work of René Descartes and Francis Bacon. The emerging modern science played a decisive role in this new world-view,[19] but it was also social acceptance of this world-view that paved the way for a positivist scientific era. Winston thus speaks of the 'social construction of technology' (SCOT) model, that also includes the social values that 'shape the precise forms that technological artifacts make'.

As theory has for long been in the spotlights of science and philosophy, the question arises what techne contributed to the symbiotic relationship. Techne, without any dispute older than theoria, can be traced back to earliest presence of man (or primates) on this planet; in the crafting of tools, clothes and all other artefacts that define the human as Homo Faber. Techne is firmly grafted in the uniquely human proficiency in tool-making and craftsmanship[20]. In techne, theoria was preceded by a fundamentally different body of knowledge , which is often termed praxis.[21]

---

[17] White, 1999

[18] Kurzweil, 2006, pp. 40-41

[19] Winston and Edelbach, 2008, p. 11, Ihde, 1993, p. 28

[20] Ambrose, 2001

[21] Ihde, 1993, p. 39

Techne and Praxis

In technoscience, theory can become disembodied from their discoverers or creators. Formal theory for instance, has to some extent means of proving theory within its own methodology. This disembodiment between discoverer and artefact also returns in praxis. Unlike theory, literature and art, the names of those who created certain artefacts are therefore often no longer known. However sad this may be for the craftspeople, it is a strength for the artefacts themselves, for they can acquire certain autonomy independent from their discoverers. I aim to argue here that within natural philosophy, the experimental method allowed theory to achieve this as well.

In order to demonstrate this, I will first introduce a pattern called a 'convergence inducing process',[22] which in my view lies at the heart of praxis and the experimental method (see next page).

This pattern, coarsely stated, consists of an actor that tests its environment in an iterative fashion and evaluates the feedback obtained from the environment by comparing it to a certain goal function. Complex environments typically have some evaluation criteria to 'certify' the test sets, which provide means for evaluation. If the actor has some kind of a securing mechanism that stores the most successful values, this process typically optimises in an exponential fashion to values that best match the goal function.[23]

To make this pattern a bit less abstract, take a classical craftsman, for instance a potter (the actor). If one realises that the lump of clay on the potter's worktable is the essential focal point in the environment of the potter then it immediately becomes clear how potter and clay co-create the eventual artefact.

The potter tests and forms the lump with her eyes and hands, while the clay gives in to, or conversely resists the force the potter exerts on the lump, and provides clues on weaknesses or rigid parts in the material. The potter may have a certain form in her mind that she aims to recreate in matter, but the clay has a say in the eventual form that is made. Borrowing Bruno Latour's notion of objects

---

[22] Alexander, 1977; Pieters, 2008

[23] Holland, 1992; Pieters, 2006

having 'agency' in a social network,[24] the convergence inducing process can now be considered a specific configuration of a socio-material actor network in which the material plays an active role.

Pattern: Convergence Inducing Process

Description *An actor samples an environment by an iterative cycle of testing and evaluation until a certain goal criterion has been met.*

a.k.a Problem Solver, Global Search

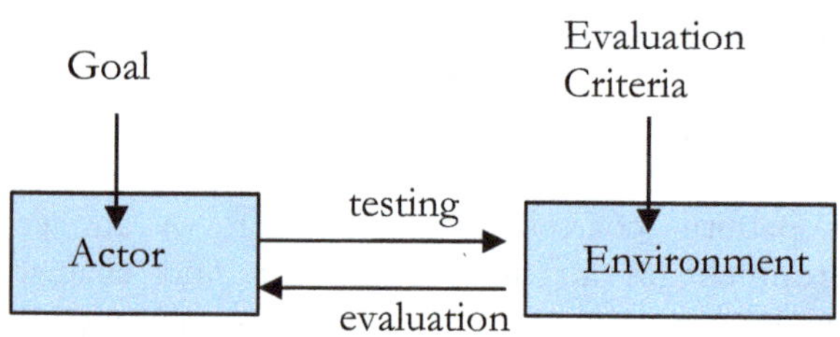

Notes *The actor typically contains a convergent process that maps the evaluated variables with the goal function*

Techne in the classic sense thus is not necessarily a Baconian submission of the material to man, but rather a communication with and through material.[25] It is a 'being-with' matter in the sense that is often heard in the stories of craftsmen and artists and, in a more modern variant of techne, expressed in Robert Pirsig's *Zen and the Art of Motor Maintenance*.[26]

It is this pattern that the experimental method introduced to theoria. Theory determines the test sets and the goals (and constraints) against which phenomena in the actors life-world are tested. The evaluation data is then used to see if theory produced sets that match that what is expected. In modern technoscience, these theories are usually (mental) models of patterns in the human life-

---

[24] Latour, 2007, p. 63
[25] Atlan, 1993, p. 39
[26] Pirsig, 1974

world. These models result in predictions that should match the expected evaluation data.
So although Bacon's claim that nature became subject to the experimentation of man can be contested from this perspective, it is clear that theory certainly did. Testing and evaluation cycles weeded – and weed- out incorrect assumptions and hypotheses. Through experiment, 'nature' is telling scientists what is possible or not, what is plausible or unlikely.[27]The classical picture of science that follows a linear path from the conception of ideas to proven hypotheses that intervene in nature does not correspond to techno-scientific practice, that more often sees an iterative approach where ideas, intuitions and hunches mature to hypotheses and theories through continuous cyclic interaction with subject matter. This is vividly represented in the development of software engineering, that initially started from a conception of a linear "waterfall model", but eventually evolved into iterative methodologies such as RAD, XTreme Programming and Agile Software Development as software became more and more complex.[28]

The feedback loop that the experimental method introduced to theory, also implies a reduction of error. In a feedback loop, error is typically reduced due to the cycles of testing and evaluation. If theory therefore produces test sets that do not match the expected evaluation data, then it is theory that needs to be readjusted. By continuously adapting theory, the mental models will yield experiments that better match the expectations. In this light, it is interesting to note that for instance Taft Broome bases his 'metaphysics of engineering' on a 'calculus of error', [29] which means that the notion of error is embedded in mathematical theory of engineering. This seems to introduce a paradox; why should one trust artefacts such as aeroplanes that are the result of models that acknowledge error? The answer again, is that this approach works because it is continuously tested and evaluated. (Theoretical) error –it is probably better to speak of mismatches with the goal function- is

[27] Fodor, 2001, p. 4
[28] Bruegge and Dutoit, 1999
[29] Broome, 2007

not necessarily a problem, as long as it is taken up in a feedback loop that can reduce this.

The experimental method forces techno-scientific theory to be continuously reflected against our natural life-world. Technological artefacts therefore also continuously reaffirm the theories that shaped them. Every car on a road confirms the laws of Newtonian mechanics, thermodynamics, physics and chemistry. Every aeroplane that successfully completes its journey reaffirms the theories behind aerodynamics and jet propulsion. Every failure that can be contributed to malfunction or improper use and not to failure of theory, reaffirms that theory.

Popper may be right in his claim that scientific theory can be falsified,[30] but in the case of established technoscience, falsification becomes more and more improbable as time goes by. It is for this reason that any paradigm shift in technoscience will usually not replace older theory, but rather change or constrain it. Einstein's theory of special relativity is, of course, a fine example of this.[31]

Embodied Technology

Technology is the art of dealing with constraints. Artefacts are produced with their own inherent complexity, which require energy, attention and resources. The constraints that these introduce, are nature's way of 'communicating' with theory. Technoscience is saturated with failed hypotheses and experiments, intuitions, wild hunches, which all serve a purpose. The iterations over these practices eventually lead to 'proven' theory and working products. Scientific articles are usually narratives in which the eventual successful route from problem to solution is recorded; where the functional dominates. However, the practice of technoscience is a far less linear socio-technical complex.

This aspect becomes even more prevalent for any activity that has a strong operational component, such as engineering or medicine, which currently are often collected under the umbrella of professionalism.[32] For professionals, the requirements of theoretical correctness experiences friction with that of having to achieve results

---

[30] Popper, 1965

[31] Einstein, 1997; Brian, 1997; Lindley, 2008

[32] Procee, 2006

in finite time with finite means. Again praxes provide the 'grease' that prevents the professional's activities from grinding to a halt. Praxes contain that knowledge that informs, amongst others, about 'probable' or 'likely' solutions, which are related to 'best-practices', heuristics, historically proven solutions and even intuitions that happen to pay off.

At this point we can see two very fundamental characteristics of technoscience. First, from a historical point of view, theory has never been crucial to the advance of techne, despite its importance for modern technoscience. Secondly, theory in the way it used in technology is very different than the way it is used in purely theoretical areas. Theory is tested and evaluated through experiment, and this defines the manner of 'reflection' of modern science; it is reflection through praxes .

Reflection on Philosophical Contributions to Technology Debates

With this, we can now return to the dawn of technoscience –of early enlightenment- and take a closer look at the historical narratives. From the perspective of a praxis engineer, these accounts are in some ways a bit confusing. They usually tell a narrative of a changing world-view in which certain names –Bacon & Bacon, Descartes, Hume, Locke, Kant and many, many more- dominate, without making clear if these persons are to be seen as spokespersons of developments that were already taking place, if they should be considered architects of an emergent Western modernity, or both.

Harry Kunneman for instance, in his criticism on contemporary Bloated Beings that fill the public domain, takes the text-book narrative of enlightenment when he considers Descartes and Bacon to be the founders of the modern world-view,[33] which implicitly assigns a role to those philosophers that goes beyond that of 'understanding through contemplation'. Apparently their philosophy has been applied in order to change our life-world. Theoria has started to pursue an agenda of change, using philosophical knowledge as an instrument. In other words, philosophy in this setting is, to some extent, technological with different means!

---

[33] Kunneman, 2005, p. 101

This, of course, is totally at par with the ideas of modernity, but it also has brought a responsibility for modern theorists in their interactions with others to make clear that their theory often aims to intervene. The German/American ethicist Hans Jonas, for instance, called ethics fundamentally action-oriented in the sense that ethics should aim to regulate human action itself.[34] For this he called on a 'heuristics of fear' to guide technological development. It is clear that such an agenda aims at some form of intervention.

From a praxis point of view this is totally understandable –praxis does this all the time- and luckily, Jonas was very clear in accounting for this. His aim was more than "understanding underlying processes"; his aim was change. But, in my view, this also excuses technologists for being wary in their interactions with for instance ethicists, especially if the interventions of the latter do not serve the agendas that technologists pursue.

From a perspective of engineering, historical narratives on technology based on theoretical analysis of texts from other theorists often give a somewhat double message. For instance, many philosophically inspired articles on the 'modern condition' of our lives are quick to point out a blaming finger to capitalistic, economic utilitarianism, scientific rationalism, and a modern enlightened world-view that proclaimed the death of God and submission of nature, while they base their criticism on the texts of other theorists who have played a role in constructing these ideas. Technologists are often on the receiving end of their criticism without having a voice in the matter. I have argued earlier that Bacon's ideas –even though they may be embraced by many modern scientists and technologists- are only one interpretation of the relationship between science and the natural, and it certainly is not one that would go uncontested to any craftsman, toolmaker, engineer or scientific experimenter that often struggles to match the ideas in her or his head with artefacts that actually do what they were intended to do. Practices often makes one humble. The voices of those who were performing the changes towards (post-) modernity are often not heard because, simply, they had other things on their mind.

---

[34] Jonas 1985, 23

Although understandable, this creates a historical void in narratives which inherently amplifies the importance of the ideas in the texts that have been made, even when these ideas are criticised.

In this light, it is noteworthy that Francis Bacon was historically fated

> … to become a background figure in early modern histories of philosophy, taking a minor role when compared to the theoretically more elegant thinkers such as René Descartes…[35]

Apparently the importance of Bacon as 'founder' of enlightenment was only appreciated in retrospect. Bacon's role has been historically amplified. It is also Ihde who warns for the risks associated with amplification. Amplification implies a:

> [...] simultaneous and necessary reduction. [...] the amplification tends to stand out, to be dramatic, while the reduction tends to be overlooked.[36]

Any production system that evolves on its own produce therefore tends to distort due to bias and effects associated with amplification/reduction.

It is therefore not surprising that it has taken female thinkers to unearth male gender-bias tendencies in the development of theory,[37] and non-Western thinkers to disclose cultural bias. Edward Said, for instance, in his criticism on modern humanist traditions, points out that his 'amphibious or bi-cultural' background offered him a 'peculiar' vantage point to observe a cultural bias in the canon of modern humanist thought and teaching.[38] However, where Said considers a continuous critical self-reflection a sufficient remedy for this risk, I would argue here that this is most likely insufficient, as it does not touch - besides possible psychological constraints- the very fundamental systemic limitations of any self-describing system that hosts processes of bias and amplification/reduction. If theory reflects through theory, then this introduces a (theory) bias in the resulting complex, self-describing production system: "[...] the linkage of

---

[35] Ihde, 1993, p. 29

[36] Ihde, 1993, p. 77

[37] Kaplan, 2003; Kunneman, 2005, pp. 70-79

[38] Said, 2004, pp. 2-3

philosophy and science might be termed theory-biased with respect to its self-interpretations."[39] With the result that blind spots stay revealed: "Theory seldom reveals its own implications."[40]

Self-reflection inherently biases that what is reflected the best. Weaknesses tend to be obliterated in the glare of that what has focus. This is the problem of any theory, because theory always implies selection. For instance, Descartes is well known for his mind over matter dichotomy, which is now often used in critique against technoscience, but which has always been prevalent in theoria rather than praxes:

> Descartes tended to dislike actual technologies and instruments because they failed to deliver the precision of deduction [...] but for that very reason much of the instrumental and technological side of actual scientific practice is overlooked.[41]

And thus, many views on technology and technologists seem to be mental constructions from theorists through theory, which are continuously reproduced, and do not necessarily have any relationship with the views that practitioners of technology themselves have. This, of course, is always the theorist's fallacy; is one aiming to understand 'reality', or is an attempt being made to ensnare reality in the straight-jacket of theory?

Towards Plurality

"All Cretans lie, says the Cretan". I turned to Toku-san in wonder. "All Cretans lie," he smiled, "or just that one, of course!"

If there is one thing that makes 'weak' theory, it is not necessarily the lack of 'strong inference' but rather a limited means of confirmation in continuous cycles of evaluation and testing against something that is outside the system. [42] The above 'solution' to the liar's paradox may

[39] Ihde, 1993, p. 22
[40] Mitcham, 1994, p. 273
[41] Ihde, 1993, p. 29
[42] Platt, 1964

be appalling for a logician, who may accuse me of not playing by the rules, but it demonstrates that many practical solutions are easy to solve if one 'jumps out of the box'.

In a world where everyone has to deal with limited knowledge, this manner of reflection 'outside the system' may be very hard, and it may even be fundamentally impossible to do so. The implicit or explicit hierarchy of norms and values that people maintain may just obscure such possibilities. Certain values, such as contempt to get one's hands dirty in practices, are injected in the production system from the start, percolates through the system and biases the eventual conclusions in a subtle kind of circular reasoning. The consequent theories therefore do not necessarily prove anything, except for making these implicit assumptions explicit, which is, of course, very important in itself. Jonas' fear was probably already present on the outset of his argumentation.

'Weak' theory does not mean that it is useless, or without meaning. In fact, any theory, except maybe those who can be proven mathematically or logically, can be considered weak at its conception. Only cycles of testing and evaluation against an external environment possibly allows certain 'certification' of theory. In this sense, theoretical methodology through criticism may just be as 'experimental' as Newton's method. Only the manner of evaluation is different.

Theorists often do not have the opportunity, the means or the will to engage in such cycles, with the result that, if one is not careful, the production of theory –weak theory that builds on other weak theory- can fan out in all directions and –for the external observer- usually ends up filling an horizon that includes the extreme oppositions and every position in-between. In retrospect, there will be always someone who is considered right –like Bacon- and thus gets historical credit. If the Industrial Era had been a disaster –which in some ways it was-, probably Jean-Jacques Rousseau would have become the founder of modernity .

Currently many –especially liberal- instantiations of modern theory seem to take a great liking to the resulting multifarious interpretation of science. Embracing the relativist nature of human understanding, they have proclaimed the death of universal truths, see a life-world that is interpreted and socially constructed , and call for many sources of inspiration to advance science. Kunneman, for one, sees allies for philosophy in the arts:

> In art and philosophy and all other explorations into new forms of expression and reflexivity the *friction between silenced possibilities* and disciplined normality continues to get a voice.[43]

Here philosophy and art are portrayed as supporting divergent processes of exploration into uncharted terrain of humanity. The lack of continuous cycles of testing and evaluation thus might actually prove to be a strength!

Although these sources may be very important in the humanities, for a lingua democratica it becomes interesting when the 'hard' sciences are summoned to be inspired by art and literature, for instance to snap out of the 'reductionist trap'.[44]

It may therefore be worthwhile to take a closer look at the systemic consequences of multifarious versus uniform approaches. In order to do so, I will draw an analogy with another pattern, which is derived from Gerald Weinberg.[45]

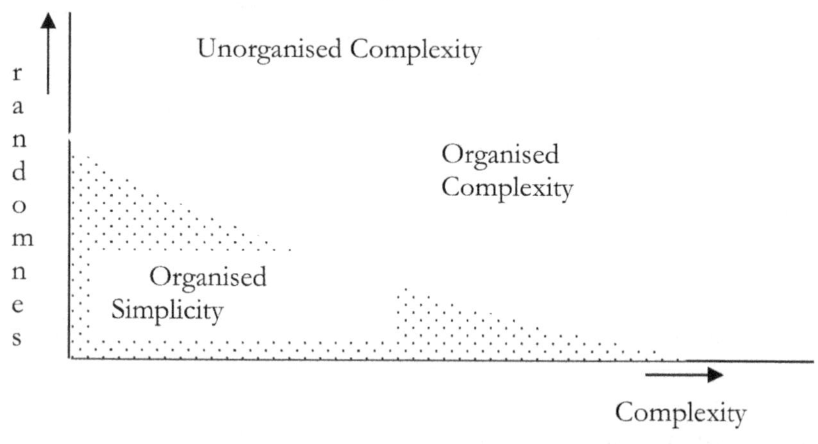

This pattern shows a relationship between 'complexity' and 'randomness' and is based on the evolution of two major scientific disciplines in physics, namely between Newtonian and statistical mechanics.

---

43 Kunneman, 2005, p. 85 (translated from Dutch by the author)
44 Atlan, 1993
45 Weinberg, 2001, p. 18, see also: Gharajedaghi, 2005, p. 92; Rasch, 1991

The first usually concentrates on systems that consist of relatively few, highly structured objects. Such systems map to the plane called 'organised simplicity' in the figure above and can usually be described best with analytical means. The other extreme consists of a large number of objects that are loosely structured (think of gas), which are best described with statistical means. This correlates to the plane called 'unorganised complexity'. This leaves one plane without means of description – organised complexity - and this is not without reason, as it is the most difficult of the threesome. In technoscience only in the last few decades the instruments and the theoretical knowledge has sufficiently matured to venture in this area. From the analytical side, we see a movement beyond causality, determinism and reductionism and other relatively straightforward means that connect theory to patterns in our life-world, while on the other chaos theory, catastrophe theory and related areas have started to unravel organisation in randomness.[46]

Now just suppose that this figure applies to the production of theory! On one hand, there is a production system that is highly structured from a small number of underlying paradigms, and which can only progress by a rigid means of proof through the rules of that system. On the other extreme, there is a multitude of highly unrelated theories with little mutual coherence. If we now accept that 'the human' in all facets is a manifestation of organised complexity, then we can understand why it was historically impossible for the humanities to align their research subject to the ideals of the logic positivist movement of the first half of the Twentieth century.[47] but also why plurality should not be cause for celebration. It may be doable for domain experts to find their way in this relativist, interpreted 'gas cloud' of theory, but this may not be expected from external observers such as technologists who, from their own traditions, domains and world-views are facing a horizon of mutually replaceable theories without any means to assess these. If these theories are also ambiguous and contradictory, the resulting spectrum of theory is totally meaningless for that observer. For that observer, a multifarious approach can thus only be dealt with through statistical

[46] Gleick, 1990; Lewin, 2000

[47] Godfrey-Smith, 2003

means. From a systemic point of view, unchecked multi-variety thus tends to skim the edges of utter meaninglessness.

Theory gas is amplified by modern demographic developments, where an emancipated, democratic and globalised academic world is flooding articles in journals and conference proceedings, where blogs, twitters and social networks spew out opinions and statements, and where art is oozing out of galleries, shopping malls and festival grounds. Everything is connected to everything in this age of information overload, causing a so-called combinatorial explosion of relationships.

In such an environment, unrestrained plurality is for unorganised complexity what reductionism is for organised simplicity; a movement away from organised complexity rather than a movement towards it.

The interested outsider who wants to incorporate the hermeneutics contained in theory gas in her or his work now faces the problem of making choices. Choices –as all selection- is the movement away from freedom; the actualisation of possibility is the end of variety. In this, pluralist theory typically offers very little assistance, tied down as it is by the systemic constraints of multi-variety. What seems to be missing is a 'philosophy of choice' that lays down beacons in the direction of this pluralist horizon.

It could be argued that this is the inevitable problem of all science. Isn't quantum physics not just as meaningless for the average human being? Well, I think not. Technology implicitly understands the art of information hiding, which selects amongst the inherent complexity of scientific production and communicates its choices through artefacts. Quantum physics reveals itself through LCD and plasma screens, mechanics and electronics through on/off buttons, touch screens, car doors and boom boxes. One does not need to know everything about technology in order to use it, and there is a whole industry of specialists who concern themselves with the issue of choice. At this level, theory is usually a background phenomenon.

The problem of choice is at the core of technology. Technology is the art of reducing imaginary forms and shapes to realised artefacts as quickly and efficiently as possible. If multifarious theory cannot help in making choices, technologists will resolve to the means they have at their disposal, for better and for worse, which is risk-analysis or other approaches that reduce these contributions to statistics. Here we see 'the End of Ethics' that Schmidt & Marratto observe in the technological age play out at systemic level, and of course statistics is also the end of philosophy! [48]

In cross-domain criticism, the humanities often tend to accuse the 'hard' or 'natural' sciences for an overly reductionist approach, and often call on 'the other' to be inspired more by philosophy, art and literature.[49] The pattern of organised complexity suggests that inappropriate reductionism is indeed present in sciences that are predominantly analytical, but a few cautionary remarks need to be made here. First, of course, the 'hard' or 'natural' sciences show more diversity than the archetype of reductionist approaches and deserve a more differentiated view. In engineering 'reductionism' is hardly a part of the domain vocabulary, as this approach is taken up in the 'game' of composition and decomposition.

More important, a critique on any activity with a call on 'plurality', 'art' or 'literature', is itself reductionist if it does not provide an analysis on how and where these sources can be embedded in the methodology or culture in which that activity takes place. A 'holistic' critique would require a certain embedding of these sources, an analysis of their strengths and weaknesses, and how and where they can be successfully applied. With respect to research in the hard or natural sciences, this careful analysis is exemplified by for instance Henri Atlan[50].

So here we are. On one extreme there is a scientific archetype that is rigid and demands proof in ways that no complex, dynamic system such as one that includes humans could ever manage to achieve. On the other extreme there are traditions that seem to be taken up in the fog of plurality, but offer few means of making informed choices for outsiders.

---

[48] Schmidt and Marratto, 2008, p. 88 and further
[49] Kaplan, 2003, Taylor, 1992
[50] Atlan, 1993

Praxes

If Weinberg's pattern holds for development of theory, then it both shows the extent and limitations of scientific archetypes, but it also explains why science has evolved in the way it did. Complex systems can never be understood through simple methodology, but simple methodology does need to be in place before it can be replaced or expanded by more complex ones. Causality, reductionism, linearity but also unrestrained plurality, maybe even 'pseudo-scientific drivel',[51] may have been needed to explore the limitations of the human intellect, but also to map the magnificent roads of its achievements.

If we accept that humanistics is rooted in pluralistic traditions, then Weinberg's pattern shows that technoscience and humanistics are divided by the space of organised complexity. However, as was argued earlier, science can also be aligned along another dimension, which goes from purely theoretical to predominantly practical.

It is here that humanistics and engineering suddenly are much closer. Engineering is firmly rooted in 'being-in' the life world. Descartes may have been right that this may be messy and dirty at times – in other words, complex - , but it also is effective and has an appealing aesthetics of its own that is still reminiscent of that of the craftsman. As was said earlier, engineering is also not as formal as even engineers themselves often like to believe. Praxes can be frustrating, tiresome, depressing but also exhilarating when achievements have been made despite the limitations of being-in-the world. Praxis means experiencing complexity, and offers unique insights that can only be seen from being in and with this complexity. It is also often not specifically taught at universities, as educators even in the praxes sciences are usually predominantly theory-focused

The same, I think, holds for humanistics. Indeed the history of the University for Humanistics itself points to a professional education of existential counselors, which positions humanistics in a tradition of 'being-with' fellow human beings that are embodied in the natural, material, social and moral world. I think this should include the

---

[51] Dawkins, 2004, p. 50

technological, functional and instrumental human being, not only as an opposition, but also as a fellow-explorer of their quests.

Humanistics, in my view, has an unique opportunity to go beyond theory in the sense that they can work from a wider network of reflection, like a mirror palace in a theme park that distorts, magnifies, reduces and reflects, but where all mirrors communicate specific patterns of communication that includes 'self' as well as the 'other' in unexpected, and therefore revealing ways. This could pave the way for a 'philosophy of choice', which moves from specific knowledge of a certain domain towards the wider (e.g. philosophical) issues that are related to that domain, through the language of that domain. Here we see a coarse outline for contextualised communications in a 'Mode-2' setting that Nowotny, Scott and Gibbons see as becoming necessary in a world where science is embedded in society.[52]

In the specific case of a lingua democratica between the humanities and technology it offers means of reflecting against science through the language of science, not only as 'criticism', but also as means of self-reflection that Said considered so important. This introduces a new way of reflection on theory beyond theory itself. Here, rather than following the safe paths of philosophical discourse or agreed conceptual frameworks, a 'factosophy', 'technosophy' or 'praxosophy' may be pursued to align domain-specific knowledge with the contributions from others.

With this, Mitcham's hope for a 'fruitful' synthesis of philosophy and technology may take an unexpected turn:

> But if technology is understood as part of a science-technology-society relationship, and philosophy takes up an active role in culture, then thinking through technology, becomes thinking through technology. [53]

If 'thinking through' is taken in the literal sense, of thinking through the means of technology itself, humanities contributions allow themselves to be 'customised' for technologists and vice versa. When this is done continuously and interactively, then we can see a 'being-with' subject matter emerge; a communication – maybe even a

---

[52] Nowotny, Scott, and Gibbons, 2001, p. 146

[53] Mitcham, 1994, p. 273

dialogue or 'horizontal transcendence' - that converges to theory that allows itself to be assessed by others.

Literature

Alexander, Christopher (1977). *A Pattern Language: Towns, Buildings, Construction.* , New York: Oxford University Press.

Atlan, Henri (1993). *Enlightenment to Enlightenment: Intercritique of Science and Myth.* New York: State University of New York Press.

Beintema, Nienke, en Anouck Vrouwe (2007). *Rond de tafel gesprekken met MCG-onderzoekers. Utrecht: NWO,* MCG-programmacommissie.

Brian, Denis (1997). *Einstein: A Life.* 1e ed. New York: Wiley.

Broome, Taft (2007). "Metaphysics of Engineering.", Workshop on Philosophy and Engineering WPE 2007, Delft.

Bruegge, Allen H. Dutoit Bernd, en Berndt Dutoit (1999). *Object-Oriented Software Engineering: Conquering Complex and Changing Systems.* Prentice Hall.

Dawkins, Richard (2004). *A Devil's Chaplain: Reflections on Hope, Lies, Science, and Love.* London: Mariner Books.

Einstein, Albert (1997). *Mijn Theorie; over de speciale en de algemene relativiteitstheorie.* Utrecht: Het spectrum,.

Fodor, Jerry (2001). *The Mind Doesn't Work That Way: The Scope and Limits of Computational Psychology.* 1e ed. Boston: The MIT Press,

Gharajedaghi, Jamshid (2005). *Systems Thinking, Second Edition: Managing Chaos and Complexity: A Platform for Designing Business Architecture.* 2e ed. Boston: Butterworth-Heinemann.

Goldberg, David E. (2007). "Why Philosophy? Why Now? Engineering Responds to the Crisis of a Creative Era." Workshop on Philosophy and Engineering WPE 2007, Delft.

Holland, John H. (1992). *Adaptation in Natural and Artificial Systems: An Introductory Analysis with Applications to Biology, Control, and Artificial Intelligence.* Boston: The MIT Press.

Ihde, Don (1993). *Philosophy of Technology: An Introduction. 1e ed.* St. Paul: Paragon House Publishers.

Jonas, Hans. 1985. *The Imperative of Responsibility: In Search of an Ethics for the Technological Age.* Chicago: University Of Chicago Press,.

Kaplan, Gisela (2003). *Gene Worship.* New York: Other Press.

Kunneman, Harry (2003). "Towards a Lingua Democratica for the Public Debate in Genomics." Proposal for a NWO Program, Utrecht.

——— (2005). *Voorbij het dikke-ik. Bouwstenen voor een kritisch humanisme.* Utrecht: Humanistics University Press.

Kurzweil, Ray (2006). *The Singularity Is Near: When Humans Transcend Biology.* New York: Penguin (Non-Classics).

Latour, Bruno (2007). *Reassembling the Social: An Introduction to Actor-Network-Theory.* New York: Oxford University Press, USA.

Lewin, Roger (2000). *Complexity: Life at the Edge of Chaos.* 1e ed., Chicago: University Of Chicago Press.

Lindley, David (2008). Uncertainty: Einstein, Heisenberg, Bohr, and the Struggle for the Soul of Science. New York: Anchor.

Mitcham, Carl (1994). *Thinking through Technology: The Path between Engineering and Philosophy*. Chicago: University Of Chicago Press.

Nowotny, Helga, Peter Scott, en Michael Gibbons (2001). *Re-Thinking Science: Knowledge and the Public in an Age of Uncertainty*. Cambridge: Polity.

NWO (2001). *De maatschappelijke component van het genomics onderzoek*. Den Haag: NOW.

Pieters, Cornelis (2006)."Effective Adaptive Plans." *International Joint Conferences on Computer, Information, Systems Sciences, and Engineering* (December 10): 277-282.

Pieters, Cornelis P. (2007). "Rethinking the Obvious. Workshop on Philosophy and Engineering", WPE 2007 (Oktober 9): 80-81.

——— (2008). "Complex Systems and Patterns. Twelfth World Multi-Conference on Systemics", *Cybernetics and Informatics* VII (Juli 26): 268-275.

Pirsig, Robert M. (1974). *Zen & the Art of Motorcycle Maintenance: An Inquiry into Values*. London: Bodley Head London.

Popper, Karl R. (1965). *The Logic of Scientific Discovery*. First Edition. New York: Harper Torchbook.

Procee, Henk (2006). *De nieuwe ingenieur*. Amstererdam: Uitgeverij Boom.

Rasch, William (1991). "Theories of Complexity, Complexities of Theory: Habermas, Luhmann, and the Study of Social Systems", *German Studies Review* 14, no. 1 (Februari): 65-83. doi:10.2307/1430154.

Rassin, Eric (2007). *Waarom ik altijd gelijk heb*. Schiedam: Scriptum Psychology.

Said, Edward W. (2004). *Humanism and Democratic Criticism*. Hampshire: Palgrave Macmillan.

Schmidt, Lawrence E., & Scott Marratto (2008). *The End of Ethics in a Technological Society*. Montreal: McGill-Queen's University Press.

Siega, Marcos. 2007. *Dexter*.

Taylor, Charles (1992). *Sources of the Self: The Making of the Modern Identity*. Harvard, Boston: University.

Weinberg, Gerald M. (2001) *An Introduction to General Systems Thinking*. 25e ed. New York: Dorset House Publishing Company, Incorporated.

White, Michael (1999). *Isaac Newton: The Last Sorcerer*. New York: Basic Books.

Winston, Morton, en Ralph Edelbach (2008). *Society, Ethics and Technology*. 4e ed. Independence: Wadsworth Publishing.

4

Ethical Complexity, Hermeneutics and Mode 3 Knowledge

Harry Kunneman

Introduction

In this chapter I want to develop a conceptual bridge between the foregoing chapters and the questions of value, meaning and politics taking central stage in the remainder of this book. In terms of the deliberative model underlying the idea of a lingua democratica for the life sciences, we could say that this lingua democratica aims to bring about dialogical connections between knowledge and insights rooted in three different domains:

1) the domain of academic disciplines involved with genomics, ranging from the life sciences and the social sciences to the humanities;
2) the domain of scientific, technological solutions and professional for concrete social and economic problems, and the divergent perspectives of different stakeholders on optimal solutions for these problems;
3) the domain of ethical values and moral frameworks, as articulated in philosophy, on the level of worldviews, in literature and art, and on the level of professional practices.

The lingua democratica for the life sciences investigated in this book, aims to provide dialogical connections between perspectives from these three domains. This is without doubt an ambitious aim. However, as transpires from the foregoing chapters, a lot of work has been done already during the last decades to clarify the conceptual issues involved. In particular the connections between the first two domains have received a lot of scientific and philosophical attention.

These connections have been conceptualized in terms of distinctions such as 'Mode 1' and 'Mode 2' knowledge;[1] the transition from "Normal Science' to 'Post-normal Science',[2] and the relation between 'descriptive' and 'normative' forms of quality control in the domain of 'science for governance'.[3] In this chapter I will focus on the third domain, the domain of ethical and moral values, and their importance for a lingua democratica. I will argue that the development of inner connections between 'descriptive' and 'normative' forms of quality control, necessitated by the emergence of Mode 2 knowledge or 'post-normal' science, asks for a specific type of knowledge and insight, which I propose to designate as 'Mode 3 knowledge'.[4] This type of knowledge is contained in ethical and moral *narratives,* articulating who we are and who we might be in relation to scientific and technological developments and all related forms of governance and professional guidance, shaping our identities. In order to distinguish Mode 3 knowledge from subjective opinions or dogmatic ideologies, we need a *criterion* for judging the ethical and moral worth of such narratives. As a first step towards the articulation of such a criterion, I turn to the idea of *ethical complexity* as developed within critical complexity thinking. I will argue that the way in which this idea is articulated by critical complexity thinkers, remains 'over-determined' by the presuppositions underlying the notion of epistemological complexity. In search for a more adequate articulation of ethical complexity, I turn to critical hermeneutics, especially the work of Paul Ricoeur. Building on his analysis of the narrative articulation of identities and on the distinction between ethical and moral questions, I try to clarify the idea of ethical complexity in terms of the ethical simplicity or ethical complexity of the 'plots' contained in the narratives told about and enacted by individuals and collectives in social practices. I suggest that ethically simple plots are connected with survival, self-preservation and competition, or 'autopoiesis' for short, whereas ethically complex plots are connected with cooperation, care and dialogical deliberation, or 'diapoiesis' for short. The development of Mode 3 knowledge, as a

---

[1] Gibbons et al. 1991; Nowotny et al. 2001

[2] Funtowicz and Ravetz, 1991

[3] Giampietro, Mayumi and Sorman, 2012

[4] Kunneman, 2005, 2010b

necessary resource for connecting descriptive and normative forms of quality care in the life sciences, can thus be understood as the effort to *enhance* the ethical complexity of the narratives underlying and legitimating the normative choices involved in the development and social circulation of knowledge and technological innovations stemming from the life sciences.

Epistemological and Ethical complexity

Critical complexity thinking is based primarily on the work of Edgar Morin, Paul Cilliers and Mario Giampietro.[5] Its main strength in the context of the search for a 'lingua democratica for the life sciences', lies in its elucidation of the inner connections between ontological complexity, epistemological complexity and ethical complexity.

On the ontological level, critical complexity thinkers assert - or imply - that the reality we humans find ourselves in consists at least in part of complex systems, characterized by emergent properties that cannot be reduced to the properties of their constituent parts. The capacity for self-organization is one of the most important emergent characteristics of complex systems. Cilliers defines this capacity as "...a property of complex systems which enables them to develop or change internal structure spontaneously and adaptively in order to cope with, or manipulate, their environment."[6] This adaptive change of internal structure, involves the development of internal models allowing both the anticipation of relevant *future* events, as Rosen says,[7] and the creation of memories. Complex adaptive systems, considered as autopoietic entities, are thus endowed with cognitive capacities that enable selections and evaluations in their interactions with their environment.[8] They actively partake in an evolutionary process characterized both by repetition and by emergence, and thus also by openness, and uncertainty.

In this way critical complexity theorists steer a middle course between materialism and idealism. On the one hand all of them are

---

[5] For a first introduction into critical complexity thinking, cf. Morin, 2008

[6] Cilliers, 1998, p. 90

[7] Rosen, 1982

[8] Maturana & Varela, 1980 (1972), p. 13

naturalists, holding fast to the material basis of evolutionary processes. As Cilliers says: "The resistance to using a reductionist discourse when describing emergent properties does not imply any resistance to materialist principles."[9] On the other hand, the core notions of emergence, self-organization and anticipatory modelling create ontological space for action, cognition and values as *descriptive* categories in the scientific modelling of complex systems.

The acknowledgement that we live in a world characterized by the evolutionary emergence of complex autopoietic systems, has far reaching epistemological implications: ontological complexity entails epistemological complexity. Critical complexity theorists along different lines articulate the meaning and import of epistemological complexity. The first line is critical and negative: it explicates epistemological complexity in terms of the fundamental *limits* of reductionist epistemologies and analytical methods. If knowledge of the properties of complex systems is based solely on knowledge of properties of their components, the properties that emerge out of the interactions between these components are made methodically invisible: "In 'cutting up' a system, the analytical method destroys what is seeks to understand."[10] In a more technical formulation inspired by Chaitin, Mario Giampietro makes essentially the same point by interpreting epistemological complexity as "the impossibility of compressing the information associated with a given perception/representation of the external world (using an algorithm) without losing relevant information."[11]

The second line along which critical complexity theorists explicate epistemological complexity underscores the *selections* that necessarily come in to play when modelling complex systems. According to Cilliers "…the knowledge gained by any description [of complex systems] is always relative to the perspective from which the description was made."[12] According to Robert Rosen "..a complex system is one which allows us to discern many subsystems depending entirely on how we choose to interact with the system."[13] In the same

[9] Cilliers, 1998, p. 93
[10] Cilliers, 1998, p. 2
[11] Giampietro, 2012
[12] Cilliers, above, p. 45
[13] Rosen, 1977, p. 229

vein Edgar Morin states: "all knowledge operates through the selection of meaningful data and the rejection of data that are not meaningful."[14] According to Giampietro

> epistemological complexity is in play every time the interests of the observer (the goal of the mapping) are affecting what the observer sees…That is, preanalytical steps - the choice of the space-time scale by which reality should be observed and the previous definition of a formal identity of what should be considered the system of interest…- are affecting the resulting numerical representation of a system's qualities.[15]

Epistemological complexity thus refers to the limits of reductionist methods and to the unavoidability of *pre-analytical choices*, as Giampietro says, determining the selection of 'relevant' subsystems, of a specific space-time scale and of specific forms of observation, resulting in selective models of the complex systems at stake.

This unavoidability of selectivity and pre-analytical choices provides the basis for a *third* line along which critical complexity-theorists explicate epistemological complexity. In the context of my argument, this line has a special importance because it provides a connection between epistemological and ethical complexity. This line concerns the influence of values, interests and relations of power in the selections and pre-analytical choices involved in the modelling of complex systems. The denial of unavoidable selectivity in the modelling of complex systems and its potentially distorting consequences by reductionist epistemologies not only leads to sloppy science. It can also serve to hide the influence of political interests and ethical values on pre-analytical problem structuring. As Giampietro says with regard to epistemological complexity in the context of sustainability:

> ..there are several applications of reductionist scientific analysis in which the problems implied by epistemological complexity can be ignored. This, however, requires acceptance without reservations from the various stakeholders who will use the scientific output of the reductionist problem structuring. Put another way,

---

14 Morin, 2008, p. 2

15 Giampietro, 2004, pp. 25-26

> reductionist science works well in all cases where power is effective for ignoring or suppressing legitimate but contrasting views on the validity of the pre-analytical problem structuring within the population of users of scientific information.[16]

The acknowledgement of the epistemological complexity at stake in the modelling of complex systems thus confronts scientists with political and ethical questions. According to Cilliers and Giampietro, these questions pertain in particular to the inclusion or oppression of different 'voices' and interests within the social system they take for granted in the pre-analytical structuring of their research problems. What narrative model of the social system they live in and what political perspective do they presuppose? According to Giampietro "power is about being able to impose your favourite narratives."[17] Against the background of his painful experiences with the Apartheids-regime, Cilliers articulates essentially the same point in the following way:

> The argument from complexity claims that a single story, or in the words of Lyotard, a 'coherent meta-narrative' cannot describe any social system fully...The reason why a certain description is acceptable has to do less with rationality and more with power. We do not have to look hard to find examples of master-narratives which oppressed the 'other' in the system, whether they be of a different race, religion, gender or sexual orientation.[18]

In this respect, the perspective of critical complexity theorists resonates strongly with the constructionist perspective elaborated by Erica Haimes in Chapter 2. In her eyes, constructionism is not "another foundational theory of knowledge but an anti-foundational dialogue".[19] According to constructionists, all claims to knowledge and the frameworks they rely upon are historically situated. Moreover they are connected with socially constituted relations of power and with social hierarchies in which some frameworks carry more weight than others on grounds which can only be justified on the basis of

---

[16] Ibidem, p. 26

[17] Giampietro, Mayumi and Sorman 2012, p. 82

[18] Cilliers, 2004, p. 24

[19] Gergen, 1995, p. 20, as cited by Haimes, above p. 62

the presuppositions of the dominant frameworks themselves. According to Haimes, the attraction of constructionism for social scientists rests precisely on this point: constructionist analyses aim to *broaden* the range of voices to be heard and try to demonstrate that there are always meaningful alternatives available for dominant viewpoints: "Constructionist analysis shows variations in meanings and diversity of sense making practices, that consequently *destabilise* views of what can be taken for granted...or as self-evident."[20]

An unresolved tension

Along these lines, the critical analyses developed by Cilliers, Giampietro, Morin, Haimes and others, shed light on the inner relations between ontological, epistemological and ethical complexity. Seen from the perspective of the development of a lingua democratica for the life sciences, this is the great merit of the conceptual framework provided by critical complexity thinking. The acknowledgement of ontological complexity implies the confrontation with epistemological complexity, which leads to the acknowledgement of selectivity and the influence of political interests and ethical values on the level of pre-analytical choices, and thus to the confrontation with ethical complexity. The confrontation with this 'double complexity' asks for systematic connections between *descriptive* and *normative* quality checks in the life sciences, as Giampietro says.

Their work thus provides an important first step towards the clarification of the ethical complexity coming into play in the development of dialogical connections between the different domains and perspectives involved in a lingua democratica. However, in my eyes the resulting analysis of ethical complexity is not radical enough. It is hampered by unresolved tensions between the general notion of self-organizing adaptive systems underlying the analysis of epistemological complexity on the one hand, and the notions of deliberation, dialogue and narrativity coming to the fore in relation to ethical complexity on the other hand. Because the notion of ethical complexity is introduced and developed in the context of a critical analysis of *epistemological* complexity, it is 'over-determined' so to

---

[20] Haimes, above, p. 76

speak by the pre-analytical choices connected with the framework of self-organizing, 'autopoietic' or complex adaptive systems. To clarify the ethical simplicity of the 'autos' presupposed in the analysis of autopoietic or complex adaptive systems, I turn to the work of John Holland.

Internal Models

In his classic introduction into the theory of complex adaptive systems, *Hidden Order,* John H. Holland lists seven basic elements shared in his eyes by all complex adaptive systems, ranging from bacteria and immune systems to mammals, companies, universities and cities. Two of these basic elements merit special attention in the context of my argument: internal models and building blocks. As underlined already with great clarity by Robert Rosen in his analysis of living beings as anticipatory systems,[21] a central characteristic of complex adaptive systems, or *cas*, is their ability to 'look ahead'. *Cas* use 'internal models', as Holland says, which allow them to anticipate and predict *future* features and events. "We usually ascribe prediction only to 'higher' mammals, rather than taking it as a property of all organisms. Still a bacterium moves in the direction of a chemical gradient, implicitly predicting that food lies in that direction."[22] Holland distinguishes between two kinds of internal models: 'tacit' and 'overt':

> A tacit internal model simply describes a current action, under an *implicit* prediction of some future desired state, as in the case of the bacterium. An overt internal model is used as a basis for *explicit,* but internal, explorations of alternatives, a process often called *lookahead.* The quintessential example of lookahead is the mental exploration of possible move sequences in chess prior to moving a piece. Both tacit and overt models are found in *cas* of all kinds – the actions and identity supplied by an immune system fall at the tacit end of the scale, whereas the internal models of agents in an economy are both tacit and overt…Despite the apparent and real differences between the bacterium's tacit model and mammalian overt

---

[21] Rosen, 1985

[22] Holland, 1994, p. 32

> models, there are important commonalities. In both cases the organism's chances of survival are enhanced by the predictions, implicit or explicit, that the model entails. Thus, variants of the model are subject to selection and progressive adaptation.[23]

Following this analysis of internal modelling, Holland introduces the notion of 'building blocks' as another, equally important basic element of all *cas*. This notion is meant to explain how internal models can help to predict relevant features of a perpetually changing environment: ".. the model can only be useful if there is some kind of repetition of the situations modelled."[24] Holland tries to explain this apparent paradox by referring to a commonplace human ability: the ability "to decompose a complex scene into parts." This decomposition is directed at basic elements or *building blocks*, which are dependable because they are "already tested for reusability by evolution and learning."[25] Holland uses the example of decomposing human faces into building blocks such as noses, ears, eyes, and hair. Each building block can be differentiated again according to some relevant characteristic such as the colour of eyes or the texture of hair. Using a limited amount of characteristic building blocks of human faces, say 10, and differentiating each of these building blocks with the help of 10 further characteristics, such as colour and texture

> we can construct any of 10^{10} = 10 billion distinct faces with these 100 building blocks. Almost any new face we encounter can be closely described by an appropriate choice from the set of 100 building blocks...This use of building blocks to generate internal models is a pervasive feature of complex adaptive systems. When the model is tacit, the process of discovering and combining the building blocks usually proceeds on an evolutionary timescale; when the model is overt, the time scale may be orders of magnitude shorter. Still...the underlying adaptive process remains much the same throughout the range of *cas*.[26]

---

[23] Holland, 1994, p. 33/34
[24] Holland, 1994, p. 34
[25] Ibidem
[26] Holland, 1994, pp. 35-37

This small sample out of Holland's broader analysis clearly illuminates the way in which agency and knowledge are conceptualized within his theory of complex adaptive systems. Holland 'borrows' his notion of agent and agency from economics. It comes as no surprise then that ethically, this notion is very simple. The 'self' implied in this model of being an agent in the world, is primarily engaged in surviving an unrelenting competition for scarce resources, by adapting successfully to a changing environment and outsmarting the competition. As a consequence, the space for ethical exploration and ethical development available to this self, is limited to two basic forms of relating to others: struggle and exchange. The same goes for the knowledge of the world gathered by this self, including knowledge of other complex systems. This knowledge has a predictive character. It is based on observation, abstraction and logical inference, leading to generalisations that allow predictions of future situations and events. So both in its relations to other entities and in its knowledge of the world, the self or 'autos' presupposed in the self-organizing activities of complex adaptive systems as analysed by Holland, is indeed ethically quite simple.

In my opinion, critical complexity thinkers like Cilliers and Giampietro remain bound up to a point to this ethically simple notion of the self and of its relations with others presupposed in the theory of complex adaptive systems. This ethically simple notion of the self shows un unresolved tension with the ethically more complex notion of the self shining through in the 'ethics of difference' proposed by Cilliers, or in the dialogical forms of political deliberation advocated by Giampietro. In the next paragraph I will argue that a more adequate explication of an ethically complex notion of the self and its relations with others, can be articulated by connecting critical complexity thinking with critical hermeneutics. With the help of concepts and ideas from critical hermeneutics it can be shown that narratives do not only serve as vehicles of power and exclusion, but also provide the medium for the articulation of ethically more complex identities and more just institutional arrangements.

Critical Hermeneutics

Critical hermeneutics provides a systematic reflection on the narrative constitution of identities, in particular the inner relation between ethics, narratives and moral rules. It is rooted in hermeneutics as an art and method of interpretation and understanding, but differs from it in important respects. Apart from its lack of attention for questions of power,[27] one of the most important drawbacks of classical hermeneutics in the context of my argument is the deep divide it has helped to create between the natural sciences and the humanities.

In the nineteenth and twentieth century, the hermeneutic craft of interpreting texts based on receptive reading, provided the background for the philosophical reflection on the methodological characteristics of the newly emerging 'Geisteswissenschaften'. Philosophers like Schleiermacher and Dilthey developed the well-known and very influential opposition between explanation and understanding, in an effort to explicate the fundamental differences between the natural sciences and the humanities. According to Dilthey we *explain* nature, but we *understand* the human.[28] In this way he tried to safeguard the methodological autonomy of the humanities against the claim that the development of causal explanations based on general laws and on controlled experiments is the only legitimate aim of scientific enquiry. This struggle has been successful up to as point in securing academic space for the humanities in which they could develop their own methods and interpretative frameworks. But it also has led to the fortification of supposedly fundamental epistemological and methodological differences between the sciences and the humanities that stand in the way of a lingua democratica for the life sciences. The unresolved tension within critical complexity thinking between an autopoietic, information processing notion of the self on the one hand and an ethically more complex conception of the self at the other hand, can be seen as a contemporary echo of these fortified differences and oppositions.

In the remainder of this paragraph I will sketch an ethically more complex notion of the self and its relations with others, building on the work of Paul Ricoeur and his analysis of the narrative figuration

---

[27] Habermas and Henrich (Eds.), 1971

[28] W. Dilthey, 1989 (1924)

of identities. On this basis it becomes possible to overcome the unresolved tension within complexity thinking by situating identities in a continuum between autopoiesis and diapoiesis. In terms of this continuum ethical complexity can be conceptualized in a more adequate and consistent way.

Oneself as another

Together with Hans-Georg Gadamer and Charles Taylor, Paul Ricoeur is justly considered as one of the most important hermeneutic philosophers of the past century. In the context of my argument I concentrate on one of his major books, *Oneself as Another*, first published in Paris in 1990. In this book Ricoeur sets himself the task to develop a notion of the identity of persons based on 'the ethical perspective', which he defines as follows: "Aiming at the good life, with and for others in just institutions."[29] This definition divides the domain of ethics into two, related fields. The first is the field of 'existential ethics', dominated by the question how to live a good life. At first sight, this question might seem to refer to individual experiences and interpretations of a good life and to individual plans and individual actions aimed at realizing a good life. However, upon closer analysis it becomes clear that *reciprocal* relations with other persons are of decisive importance for living a good life. Following Aristotle, Ricoeur refers to genuine friendship as a paradigmatic example of an interpersonal relation that contributes to the realization of a good life for *both* friends. This relation brings out 'the best' in both of them: friends enable each other not only to experience 'goodness' by receiving it from the other, but also by *giving* it to the other, in the form of attentiveness and care, or 'solicitude', as Ricoeur says.

The mutual solicitude that characterizes genuine friendships connects the field of existential ethics with the second ethical field: *moral* justness on the level of institutions. This field of 'institutional morality' brings in another ethical criterion, which presupposes the criterion of a good life, but adds the idea of an equal and just *distribution* of the resources needed for a good life. According to Ricoeur the concept of distribution must not be limited to the

---

[29] Ricoeur, 1992 (1990), pp. 172/180

economic plane, as a complement to production: "It denotes a feature fundamental to all institutions, to the extent that they govern the apportionment of roles, tasks, and advantages or disadvantages between the members of society."[30] According to Ricoeur, equality is to life on the level of institutions what solicitude is on the level of interpersonal relations. On the level of interpersonal relations, we have another person before us who has a 'face'.[31] In contrast, equality on the level of institutions is applicable to *all.* It opens the moral horizon of universal justice.

Five components of an ethically complex notion of the self

Against this background I can now provide a (highly stylized) sketch of the ethically complex image of the self and its relations with others that can be gleaned from Ricoeur's work. My sketch of this ethically complex image of the self is build up out of five components.

1. The *first* component of this ethically complex image of the self rests on the distinction between two different poles of the identity of persons, two different ways of being the 'same' person. Ricoeur designates these two poles with the Latin words *idem* and *ipse.* The identity of persons as *idem* concerns their 'character', understood as their characteristic way of being in the world. It refers to all that remains the same in the flux of time and makes them recognizable for others. The idem-pole of identity thus refers to sameness as a specific form of permanence in time. In terms of the theory of complex adaptive systems, we could say that the idem-pole of the identity of persons can be observed and described in terms of unchanging 'building blocks' of their identity: habitual ways of behaving, characteristic ways of expressing; in short all that can be *predicted* about ourselves with a certain measure of confidence.
The description of the identity of persons as *idem,* thus answers the question *what* we are. In contrast, the ipse-pole of personal identities cannot be described in this way, because it concerns the question *who* we are. According to Ricoeur, answers to this question refer to a

[30] Ibidem, p. 200

[31] Ricoeur refers here to the ethics of Emanuel Levinas. Cf. f.i. Levinas, 1969 (1961)

different form of permanence in time. He gives two paradigmatic examples of the ipse-pole of identity: on the one hand keeping one's promise and being dependable for another who counts on one's solicitude and care; on the other hand standing up for and being true to one's moral convictions, also designated by him as 'attestation'. In both paradigmatic examples the specific form of permanence in time characterizing the ipse-pole of identity, emerges in *answer* to the needs and hopes of another person (or persons). This emergence is precarious. In many cases, this ethical form of permanence in time collapses under the pressure of the idem-pole of identity; that is, under the pressure exerted by specific 'building blocks' of ones character whose weight pushes one in other directions then being faithful to ones promises or standing up for equality and justice.

2. At this point the *second* component of the ethically complex image of the self provided by Ricoeur comes in: narrative self-modelling, or 'narrative identity', as Ricoeur says.[32] According to him, the tension between *idem* and *ipse*, or between sameness and selfhood, is 'mediated' by the *narrative* construction and reconstruction of the unity of one's life. In this way Ricoeur radicalizes the hermeneutic tradition, by shifting the focus from the interpretation and understanding of texts and narratives, to the interpretation and understanding of oneself! According to Ricoeur we can and do consider ourselves as the main character in our life story. This story develops by way of an expanding network of small stories about our experiences, feelings, hopes and fears. Not only the small stories we tell to ourselves and to others, but also the stories told by others about us. Our identity is shaped in a dynamic network of interlocking stories as a *narrative* identity, as an on-going process of narrative self-modelling. In this process, the tension between the idem-pole and the ipse-pole of our identity is mediated by narrative figuration and re-figuration. On the one hand we confirm and stabilize the idem-pole of our identity by way of little stories about our selves in which our character is confirmed and stabilized. On the other hand, the appeal of others to our solicitude can motivate us to re-imagine and re-configure up to a point the fixed mould of our identity, in so far as it impedes us to answer the appeal of others, or hinders us be true to

[32] Ricoeur, 1992, pp. 140-168

our deepest moral convictions. This appeal can help us to imagine another, ethically more complex future continuation of our life story and act accordingly. In *answer* to the question *who* we are, posed to us by others who depend on our solicitude, we can change in small or even in decisive respects the content of our life story, and thus to some extent of our identity. The plot of our life story, and thus of our character as its main protagonist, becomes ethically more complex. The question "Who am I, so inconstant, that *notwithstanding* you count on me?" can be answered by stating: "Here is where I stand."[33]

3. The *third* component of the ethically more complex image of the self provided by Ricoeur, is closely related to the first two, but brings in another level: the level of *practices* in which persons actively participate to bring about some good, such as good education, good care, or a good pop band. The notion of character and the dynamic relation between idem and ipse underline the *individuation* that is internally connected with and even constitutive of an ethically complex notion of the self. The introduction of practices as a crucial component brings in the equally constitutive role of *cooperation* in an ethically complex notion of the self. In Dutch, my native language, cooperation reads as 'samenwerking', akin to the German noun 'Zusammenarbeit'. In both languages cooperation is literally designated as *working together*. The combination of these two elements, 'together' and 'work', helps to elucidate the nature and role of practices as the third component of an ethically complex image of the self.

To start with practices are constituted by their relational character. "Practices are based on actions in which an agent takes into account, *as a matter of principle*, the actions of others," as Ricoeur says.[34] This 'taking into account' not only concerns the cooperation that provides the basis for all practices, from small households to large orchestras. Practices moreover provide ideals and standards for 'good work', for bringing about a 'good' in the world on a cooperative basis. The degree in which these ideals and standards are met is judged *both* by those who perform the work and try to bring about the goods central

[33] Ricoeur, 1992, p. 168
[34] Ricoeur 1992, p. 155 (Italics added, HK)

to a specific practice, *and* by those who make use of -or are dependent on - this work for the realization of a good life.

Although Ricoeur does not explicitly draw this conclusion himself, in my view this perspective on the role of practices points to a third articulation of the ipse-pole of identities, besides being faithful to others in response to their need for our solicitude and standing up for and being true to one's deepest moral convictions. This third articulation of the ipse-pole of identities concerns answering a 'professional calling', that is: dedicating oneself to realizing as a professional *practitioner* the ideals and standards of specific practices and develop them further. Along this path, the narrative of one's life becomes enmeshed not only with the narratives of the lives of concrete others to whom one is related as friend, partner, father, daughter etcetera. Moreover it becomes enmeshed with the broader narrative of specific practices and their continuing history.

In my eyes, this continuation partakes in the same temporal complexity that Ricoeur has uncovered in the dynamic relation between *idem* and *ipse*. This continuation forms a dynamic mixture of two different forms of permanence in time: on the one hand repetition of the same, on the other hand commitment to – and trust in - the *future* realization of a good life 'with and for others'. Trying to hold fast to ones promise to 'be there' for others, struggling to be true to ones convictions and following a professional calling, depend all three to a certain extent on an *imagined,* narratively constructed *future* realization of a good life. The realization of this imagined good life in practice, on the level of concrete actions and concrete forms of cooperative work, is both dependent on and stands in tension with repetitive patterns and mechanisms.

4. The *fourth* component of an ethically complex notion of the self that I glean from his work concerns the last part of his definition of the ethical aim: 'a good life, with and for others *in just institutions'*. As indicated already above, Ricoeur explicates the idea of justice at the level of institutions with the help of the notion of equality, more in particular the equal distribution of the 'goods' and 'burdens' connected with living together and working together in the framework of institutions. In his eyes, the notions of equality and justice bring in a new perspective compared with the personal perspective of 'I' and the interpersonal perspective of 'You' that come into play in the realization of 'a good life, with and for others'.

Over and above the perspective of *concrete* others, the ideas of equality and justice bring in the perspective of others whom we will never meet face-to-face, the perspective of 'Each' as Ricoeur says.

> Equality provides to the self another who is an *each*...the sense of justice takes nothing away from solicitude; the sense of justice presupposes it, to the extent that it holds persons to be irreplaceable. Justice in turn adds to solicitude, to the extent that the field of application of equality is all mankind.[35]

Here we enter the domain of *universal* moral values and moral precepts, such as the Golden Rule, Kant's categorical imperative and Cillier's 'provisional imperative', outlined in Chapter 1.

At this point in his argument Ricoeur brings in the idea of 'evil'. In his eyes, universal norms of justice are necessary as a kind of 'sieve' that can sort out self-centred articulations of the good life and can serve as a check against the 'propensity for evil' of human beings. This propensity refers not only to all forms of *indifference* to harm inflicted upon people by 'ill fate' and by the actions of others, but also to the capacity of humans to actively *inflict* harm on others, both directly, by acts of violence, and indirectly, by taking 'more then their just share' from the goods flowing from cooperating in institutions. Universal moral values and moral precepts are necessary in his eyes to check the different manifestations of this 'propensity for evil' and gradually diminish their role in human life.

5. Universal norms of justice are a necessary condition for just institutions, but they are not sufficient. The effort to heed universal norms in practical situations brings a *fifth* component of an ethically complex self into play. Building on Aristotle's notion of 'phronesis', Ricoeur designates this fifth component as practical wisdom. In many practical situations, the precepts contained in universal norms are too general to do justice to their complexity. According to Ricoeur in such situations we have to reintroduce the ideas of a good life and of solicitude for concrete others, in order to determine an appropriate and 'wise' course of action.

---

[35] Ricoeur, 1992, p. 202

In this way Ricoeur forges a fruitful connection between the Aristotelian tradition focusing on the good life on the one hand and the Kantian tradition focusing on universal norms on the other hand. Universal moral norms are needed as a 'sieve' for articulations of a good life implying indifference or injustice with regard to other human beings. But confronted with complex practical situations we have to bring in the ethical perspective of a good life with and for others again, in order to be able to act wisely.

Autopoiesis and Diapoiesis

Along these lines, Ricoeur's work provides important conceptual building blocks for a notion of the self and its relations with others that is ethically much more complex than the autopoietic notion of the self presupposed in critical complexity thinking. As argued above, the notion of ethical complexity is originally developed in the context of a critical analysis of *epistemological* complexity. As a consequence it remains 'over-determined' by the pre-analytical choices connected with the framework of self-organizing, complex adaptive systems. With the help of the ethical complex notion of the self derived from Ricoeur's analysis of narrative identities, it becomes possible to free the crucial notion of ethical complexity from this 'overdetermination' by the ontological and epistemological presuppositions connected with the idea of autopoietic systems.

To clarify these implications, it is helpful to compare the ontological and epistemological presuppositions of an analysis of complexity connected with an ethically simple, autopoietic notion of the self, with the presuppositions of an analysis connected with an ethically complex, or 'diapoietic' notion of the self (see next page). The term 'diapoiesis' is meant as the opposite pole of autopoiesis on a scale of increasing ethical complexity. Terminologically it is based on the combination of 'dia' - from 'dialogue' and 'dialogical' - with the notion of 'poiesis', as articulated by Maturana and Varela in their analysis of autopoiesis. It refers to a broad range of ethical and moral relations with others - from solicitude and care, to cooperation and dialogical deliberation - that can emerge out of the tension between the idem-pole and the ipse-pole of identities flowing from the appeal of others for our solicitude, the appeal of a professional calling or the appeal for more just institutions.[36]

When ethical simplicity and ethical complexity are analysed in terms of the differences between an autopoietic and a dia-poietic notion of the self, it becomes clear that the ethical simplicity of the ontological and epistemological framework presupposed by the idea of autopoiesis rests on two related pre-analytical choices.

Articulations of complexity	Based on an autopoietic notion of the self	Based on a dia-poietic notion of the self
Ontological complexity	The world shows emergent properties and contains self-organizing systems anticipating future states of their world with the help of internal models, and evaluating their experiences and actions in view of their own survival and wellbeing	The world contains other living beings like myself, trying to live a good life. My relations with them can be based on competition for scarce resources, but also on answering their appeal for cooperation and care
Epistemological complexity	All models of complex systems are based in part on pre-analytical, normative choices influenced by dominant narratives and relations of power	I am always more than can be revealed by descriptions of what I am. In answer to the appeal of others who need me, I can reconfigure my idem-identity
Ethical complexity	Because asymmetrical relations of power influence pre-analytical choices in the modelling of complex systems, descriptive and normative quality checks have to be combined by broadening the range of stakeholders and strengthen the countervailing power of their voices	Individual articulations of a good life should pass the 'sieve' of universal norms, in view of just institutions; but these norms do not provide a final basis for just action. Practical wisdom is required in concrete situations

[36] Cf. H. Kunneman, 2010a

In the first place, the *appeal* of others to our solicitude, or the appeal for a more just distribution of essential resources- and our capacity to answer such an appeal are absent from the analysis. As a result the *tension* between the idem-pole and the ipse-pole of identities, which is constitutive for an ethically complex notion of the self, is absent too. In the second place, the internal models of self and the world presupposed in the framework of autopoiesis are restricted to descriptive, predictive and explanatory models. The specific potential of *narrative* models of the self and the world to *imagine* an ethically more complex identity and to *inspire* a commitment to more just institutions, is lacking. Narrative models of the self and the world are not only descriptive and predictive but also narrative and figurative. They are constructed around narrative plots ranging from autopoietic self-preservation, competition and cooperation, to care, dialogue and furthering just institutions. When the imaginative and inspiring character of narratives is connected with Ricoeur's idea of the narrative reconfiguration of identities, it becomes clear that 'what' we are does not determine in a linear way ' who' we are. In answer to an ethical or moral appeal a narrative re-figuration of our identity can emerge, resulting in another pathway at crucial bifurcations in the dynamic development of our life story. The relation between what we are and who we are is non-linear.

It is clear however that ethical and moral relations connected with ethically complex identities remain fragile. Their ethical complexity is connected with *engaging* the tension between the idem-pole and the ipse-pole of one's identity. Bifurcations potentially leading to autopoietic forms of relating to oneself and to others keep emerging. People can and frequently do break promises and give up on care and solicitude, they can unwillingly or intentionally deeply damage others and they can succumb to the seduction to take more than their fair share of jointly created resources for a good life, or become deaf to the appeal for more just institutions. But we can also *learn* in this crucial respect. We can learn to interpret our own identity with the help of ethically more complex plots and we can deepen our practical wisdom enabling us to connect universal norms of justice with the needs and possibilities of concrete persons in practical situations. As I will argue in the concluding paragraph, this possibility is of great importance for the development of a lingua democratica for the life sciences. Can we develop ethically more complex narratives with regard to the influence of the life-sciences on our identities and our relations with others?

Conclusion: the Importance of Mode 3 Knowledge

In the coming decades the new possibilities for influencing and manipulating our genetic characteristics provided by the life sciences, will not only heavily influence *what* we are but will also increasingly shape possible answers to the question *who* we are. Clear illustrations of this influence (further analysed below) are provided by expected future possibilities for individuals to acquire extensive knowledge with regard to possible health risks early in their lives, or to lengthen considerably the span of their life histories, or to influence genetic characteristics of their children and grandchildren according to personal preferences, or in accordance with socially desirable characteristics. As critical complexity thinkers and social constructivists have argued, the normative choices at stake here are not added 'from outside' to independent or autonomous scientific and technological developments. They are embedded within science and technology themselves by way of the pre-analytical choices guiding the construction of scientific models and by way of taken for granted norms and values in the professional practices connected with them.

Concepts such as Mode 2 knowledge and 'post-normal' science are meant to do justice to the increasing influence of such normative choices on the development of the life sciences. In conclusion I want to argue that confronting these normative choices in a democratic and responsible way, asks for a specific type of knowledge and insight, which I designate as 'Mode 3 knowledge'.[37] This type of knowledge has a *narrative* character. It is contained in ethical and moral narratives, articulating *who* we are and who we might be *in relation to* scientific and technological developments and in relation to all forms of governance and professional guidance connected with them. However, as argued already in the introduction, to judge the ethical and moral worth of such narratives, a criterion is needed in

[37] Kunneman, 2005, 2010b

order to distinguish valid Mode 3 insights from subjective opinions or dogmatic ideologies.[38]

Against the background of my analysis of ethical complexity with the help of Ricoeur's hermeneutic perspective, I suggest that the ethical complexity of the plots contained in narratives, articulating *who* we are and who we might be *in relation to* scientific and technological developments, can be used as a criterion for the validity of Mode 3 knowledge. In ethically simple plots, the question 'who am I?', as an ethical and moral question, does not even come to the fore. It remains 'covered up' so to speak by descriptive answers to the question what we are, such as descriptions of human beings as part of 'nature red in tooth and claw', or as complex adaptive systems "able to develop or change internal structure spontaneously and adaptively in order to cope with, or manipulate their environment."[39] When narrative plots move away from the pole of autopoiesis in the direction of the pole of diapoiesis, ethically more complex plots come to the fore, such as cooperation and care. The decisive step towards the pole of autopoiesis is taken however, when such ethically more complex relations towards other living beings are articulated not only as properties of the actors involved that can be *described* as part of a narrative plot, but also as the emergence of a new plot at the level of the narrative identity of the actors themselves, that is to say: as a specific answer to the question *who* they are. Such an answer combines an answer at the level of the identity of the person involved – who am I? - and an answer *for another person*: who are you *for me* when I appeal to you to help me or care for me? The ethical complexity of such an answer hinges on the emergence of a diapoietical relation out of the *tension* between the idem-identity and the ipse-identity of the person involved, in answer to an appeal from another person for help or care. At the pole of autopoiesis, this tension is absent. At the other, dia-poietic pole however it is very high, because at this pole narrative plots articulate forms of practical wisdom based on the effort to heed *universal* norms of justice in concrete situations.

---

[38] See on this point also Marli Huijer's analysis in the final chapter of this book.

[39] Cilliers, 1998, p. 90

Along these lines a criterion could be constructed for judging the ethical and moral worth of narratives articulating who we are and who we might be in relation to scientific and technological developments and in relation to the different forms of governance and professional guidance connected with them. Such a criterion could be of considerable importance for the further elaboration of the idea of a lingua democratica for the life sciences. As I stated in the introduction to this chapter, this lingua democratica aims to bring about dialogical connections between knowledge and insights rooted in three different domains:

1) the domain of academic disciplines involved with genomics, ranging from the life sciences and the social sciences to the humanities;
2) the domain of practical scientific and technological solutions for concrete social and economic problems connected with specific professional practices, and the divergent perspectives of different stakeholders with regard to optimal solutions for these problems;
3) the domain of ethical values and moral frameworks, as articulated in philosophy, on the level of worldviews, in literature and art, and on the level of existential ethics.

The lingua democratica for the life sciences aims to bring about dialogical relations between perspectives from these different domains. Dialogical relations however ask for a certain measure of *equality* between the different participants and their perspectives. Seen from a traditional, Mode 1 or 'normal science' perspective, the diverging interests of different stakeholders coming into play in the domain of Mode 2 knowledge or in 'post-normal' science, appear as a contamination of the objectivity of scientific knowledge. This goes even more for ethical values and moral frameworks. Seen from the perspective of critical complexity thinking and social constructionism, normative choices cannot be avoided in the modelling of complex systems. However, their 'advice' to further dialogical forms of deliberation between the normative perspectives of the different stakeholders involved, presupposes that the problem they address is already solved. Why would these different stakeholders be *prepared* to enter a dialogue, broaden the range of voices to be heard and consider meaningful alternatives for their own perspective?

This question obviously is of great importance for the development of a lingua democratica for the life sciences. With the help of Ricoeur's analysis of narrative identities it can be reformulated in terms of the tension between the idem-pole and the ipse-pole of identities and the development of ethically more complex identities. According to Ricoeur, moral commitment to social justice is rooted in the existential ethics of individuals. The development of ethically more complex identities can emerge out of the *appeal* for help and care, out of a professional calling and out of the appeal for more just institutions. Such appeals can help to reconfigure our identities up to a point and narratively imagine a different future continuation of our life history. For such narrative reconfigurations at the level of personal identities, the imaginative and inspiring potential of narratives at the level of specific practices and at the level of the wider culture is of decisive importance. The idea of Mode 3 knowledge tries to address this need with regard to the domain of practical scientific and technological solutions for concrete social and economic problems and the professional practices connected with them. In view of the increasing influence exerted by the sciences in general, and by the life sciences in particular, not only on what we are, but also on who we are, the idea of Mode 3 knowledge points to the great importance of learning to interpret our technologically mediated identities with the help of ethically more complex narrative plots in the context of the professional practices in which this mediation is brought about. A lingua democratica for the life sciences can be developed, by furthering dialogical connections between three different, but equally important Modes of knowledge.

Literature

Cilliers, P. (1998). *Complexity and Postmodernism.* New York: Routledge

Cilliers, P. (2004). *"Complexity, ethics and justice." Journal for Humanistics,* 5(19), 19-26.

Dilthey W. (1989) (1924). *Introduction to the Human Sciences.* Selected Works, Vol. 1, R.A. Makkreel and F. Rodi (Eds.), Princeton: Princeton University Press

Funtowicz, S.O. and Jerome R. Ravetz. (1991), "A New Scientific Methodology for Global Environmental Issues." In: *Ecological Economics: The Science and Management of Sustainability,* ed. Robert Costanza. New York: Columbia University Press, pp.137-152.

Gadamer, H. G. (1989) (1968). *Truth and method (2 rev. edtition ed.).* New York: Crossroad.

Giampietro, M. , (2004.) *Multi-Scale Analysis of Agro-ecosystems.* Boca Raton/ London, CRC Press

Giampietro, M., Mayumi, K. and Sorman, A. (2011). *The Metabolic Pattern of Societies: Where Economists Fall Short*, New York: Routledge

Habermas, J. and Henrich, D. (Eds.). (1971). *Hermeneutik und Ideologiekritik.* Frankfurt a.Main: Suhrkamp Verlag.

Haimes, E. and Williams, R. (1998). *"Social constructionism and the new technologies of reproduction." In Velody, I. and Williams, R. (Ed.), I (pp. 132-146).* London: Sage.

Holland, John H. (1994), *Hidden Order.* New York: Helix Books

Kunneman, H. (2005). *Voorbij het dikke-ik.* Amsterdam: SWP/Humanistics University Press.

Kunneman, H. (2010a). *"Ethical complexity." In Cilliers, P. and Preiser, R. (Ed.), Complexity, difference and identity.* Dordrecht/Boston: Springer, pp.135-172.

Kunneman, H. (2010b). *"Viable alternatives for commercialized science: The case of humanistics." In H. Radder (Ed.), The commodification of scientific research. Science and the modern university.* Pittsburgh: University of Pittsburgh Press, pp.307-336).

Levinas, E. (1969) (1961). *Totality and Infinity: An Essay on Exteriority.* Pittsburgh: Duquesne University Press

Maturana, H.R. and Varela, F.J. (1980). *Autopoiesis and cognition : The realization of the living* . Dordrecht/Boston: D. Reidel Publishing Company.

Nowotny, H., Scott, P. K., & Gibbons, M. (2001). *Re-thinking science: Knowledge and the public in an age of uncertainty* . Cambridge, UK: Polity.

Ricoeur, P. (1984). *Time and narrative, vol. 1.* Chicago: The University of Chicago Press

Ricoeur, P. (1992) (1990) *Oneself as Another.* Chicago: The University of Chicago Press

Rosen, R. (1985). *Anticipatory Systems: Philosophical, Mathematical and Methodological Foundations.* London: Pergamon Press

Ulanowicz, R. E. (1997). *Ecology, the Ascendent Perspective.* New York: Columbia University Press.

Part 2

Ethics and Worldviews

5

Morality and Genomics: Which Comes First? Towards Cooperative Inquiry

Cor van der Weele

Patterns of attention in ethics

In the Netherlands, people are so used to living below sea level that we do not experience it as special; we simply take it for granted. Only on special occasions, for example when a guest from abroad shivers at the thought of walking "on the bottom of a lake" with the North Sea above our heads, we Dutch will pause for a moment to realize how extremely artificial this country is.[1]

Human attention is selective, and the selectivity is guided by all kinds of more or less automatic mechanisms. Our attention is drawn to what is new and unusual rather than what is familiar, to what is conspicuous rather than inconspicuous, to what is present rather than absent, and so on. These mechanisms usefully alert us to changes in our environments, which may represent opportunity or danger, but they also render our attention vulnerable to hypes and manipulation, and cause us to overlook the value and importance of stable features of our environment. Discovering the significance of the familiar often requires that we distance ourselves from it, so that we can see it with different eyes. The eyes of strangers can be effective, reminding the Dutch how strange their country is. Threats, doubts and calamities also have the power to disrupt the self-evidences of daily life. Our national water managers reportedly pray for a little calamity every now and then, to remain assured of the goodwill and tax money they need in order to maintain our dikes, dams and pumping stations. Sea level rise will do, too: if it continues, all the Dutch will sooner or later attend to the lowness of the land.

Patterns of attention in science and ethics are morally important and therefore deserve attention in their own right, which they increasingly

---

[1] Ward, 2002, pp. 16-33; Van der Weele, 2004, pp. 15-16

receive, for example in the form of reflection on the research agendas of genomics and ELSI-genomics.[2] In a previous paper, I wondered what would be interesting to know about ELSI-genomics agendas, and how to achieve such knowledge.[3] Apart from patterns and clusters, it is also important to understand these patterns, as well as their lacunas, I argued. Empirical inventories of ELSI research give us patterns, but they are not so good in finding blind spots and guiding thoughts.[4] For that purpose you need perspectives on the agenda-setting generative ideas, systematic ways of looking from a distance. In an attempt to contribute tools for that job, I distinguished three different guiding ideas for (three different) moral agendas for genomics. The ELSI research agenda is characterized by the idea that society builds on science, while two alternative agendas are generated by ideals of democratization and global justice, respectively. Blind spots in one agenda can be discovered from the point of view of the others. The paper therefore argued that a plurality of agenda-setting approaches is needed. It also argued that the global justice perspective needs more attention.

The present chapter takes up the same issue, patterns of attention in moral agendas for genomics, and continues the search for understanding through a modification and simplification of the distinctions. I will distinguish two alternative sources from which the main agenda of bioethics can be derived: science and the world as a whole. If science is the source of the bioethical agenda, ethics follows science. This is a pattern with which we have become extremely familiar, as bioethics is often defined through it. For example, the Dutch Society for Bioethics explains on its website that bioethics "concentrates on moral questions which arise from developments in the so called life sciences".[5] But ethics can also derive its main agenda from an analysis of the state of the world. In the following two paragraphs, I will take a closer look at these different ways of agenda-

---

[2] ELSI: Ethical, Legal and Social Implications

[3] Van der Weele, 2008

[4] I argued this based on two inventories: Wouters, 2005 and Avard et al. 2006.

[5] To be found at www.nvbe.nl. Another example is the Wikipedia-lemma on bioethics. The first sentence says: "Bioethics is the study of controversial ethics brought about by advances in biology and medicine." http://en.wikipedia.org/wiki/Bioethics

setting. After that, I will propose that recognition of these different starting points need not inevitably result in a dispute of priority, but can also lead to cooperation.

Starting from genomics

New technology has a prominent place in the theory and practice of innovation, even to the extent that it may seem self-evident that innovation is associated with developments on the frontiers of science and technology. Various images serve to illustrate this primacy of science and technology. The frontier metaphor is one of them, with a strong significance especially in the United States. There is probably no image that surpasses its mythical power in capturing the spirit and identity of the USA. The frontier hypothesis, proposed by Turner in 1893, suggests that the great westward expansion, a journey into the unknown, captured the American idea of fresh beginnings and shaking off old European habits. The frontier produced democracy American-style, rooted in individualism. And it is to the frontier, Turner said, that the American intellect owes it striking characteristics:

> That coarseness and strength combined with acuteness and inquisitiveness; that practical, inventive turn of mind, quick to find expedients; that masterful grasp of material things, lacking in the artistic but powerful to effect great ends, that restless, nervous energy, that dominant individualism, working for good and for evil, and withal that buoyance and exuberance that comes with freedom.[6]

The "frontier hypothesis", or myth, or metaphor, is archetypically heroic and continues to inspire. It inspires scientific and entrepreneurial endeavors as well as their combination.[7]

Genomics is one of the great areas of frontier activity, and the use of the frontier imagery in the context of genomics has often been noted. In a paper on metaphors of morality in the Human Genome Project, for example, Alice Dreger noted that, first, the Human Genome Project is seen as part of the great American tradition of

---

[6] Turner, 1893, Chapter 1

[7] Many examples can be found in Slotkin, 1992

exploration and settlement.[8] Second, and more specifically, the effort to chart genetic mysteries is regularly compared to the exploratory journey of Lewis and Clark and the opening up of the American West. When in 2000 the draft of the Human Genome was presented at the White House, the Lewis and Clark map even was the centre of an almost religious adoration, according to an analysis of Hub Zwart.[9]

Other metaphors and images also express the primacy of science and technology, specifically genomics, for human progress. Science is often thought to lay new foundations for society to build on, and in "the genomic era" this image is very much alive. An influential agenda-setting paper by Francis Collins and co-authors[10] presents genomics in this way: it is the foundation for a new societal building. On the basis of this metaphor, illustrated by a carefully drawn image, the text sets out to describe an ambitious research program. The human genome project is pictured in a sphere of great hope and aspiration: "If we, like bold architects, can design and build this unprecedented and noble structure, resting on the firm bedrock foundation of the HGP, then the true promise of genomics research for benefiting humankind can be realized."[11] (see next page)

In both these metaphorical views of societal progress, the role of science is primary and that of ethics secondary. In the image of the building, genomics grounds a new society. Ethics, as part of ELSI (Ethical, legal and social implications), has a place as one of the vertical crosscutting structures of the building that serve to connect the different floors (which stand for societal themes). They deal with the consequences of genomics and help society "to capitalize on the immense potential of the HGP to improve human health and well-being."[12] The paper notes that ELSI has already resulted in substantial increases of public awareness and in significant but still incomplete protections against misuses such as genetic discrimination.[13]

---

[8] Dreger, 2000
[9] Zwart, 2009, p. 31
[10] Collins et al. 2003
[11] Collins et al. 2003, p. 847
[12] Collins et al. 2003, p. 836
[13] Collins et al. 2003, p. 835

The future of genomics rests on the foundation of the Human Genome Project.[14]

In the frontier imagery, too, ethics is an important but secondary activity. At the historical American frontier, civilizing activities and rounding off rough edges were predominantly women's work, according to the analysis of Julie Jeffrey in her book Frontier Women.[15] In contrast to what is sometimes thought, she writes, the social roles of men and women, which had become separate when the house was no longer the place where money was earned, remained separate on the frontier. The world of men was and remained outside, public, competitive and dangerous. Women's sphere was the home, where they provided not only safety and tranquillity, but also defended values and protected the vulnerable family against the bold harshness of the frontier world.

In the context of genomics, the role of the ELSI program is comparable to the role of women at the frontier. ELSI takes the developments of genomic science as a starting point, wonders about

---

14 Source: Collins et al. 2003. Courtesy of the National Human Genome Research Institute

15 Jeffrey, 1998

the ethical, legal and social implications, and guards over morality. As the explanation of the building also confirms, ELSI civilizes and protects those who are vulnerable.

Bioethics in general can be argued to have this protective societal role with regard to the life sciences, as I noted in the introduction; it typically defines itself as dealing with the implications of advancements in the life sciences. Principles, norms and values are its guiding tools.

The role of ethics in these metaphorical views of progress is not only secondary from the point of view of progress or innovation. It is secondary even from a moral point of view, since the primary moral activities are science's expansive frontier bravery and/or foundational activities. The pioneering hero is the researcher who enables new and undreamt-of improvements of human existence. His wife ELSI will take care of civilizing the technology, smoothing the rough edges. In this model of the relation between science and society, science takes the lead.[16]

The primacy of science and technology has led to a division of roles in debates on new technologies: technologists sketch the societal promises and potential benefits of the new technology, while ethicists subsequently argue that this framing is too narrow and that it overlooks societal and moral problems. Ethics has thus become associated with a critical role and a focus on caution against risks. Swierstra and Rip analyzed moral argumentation about new and emerging technologies along these lines.[17] The pattern they describe starts when a new technology full of options and promises is announced. In response, people pleading for prudence and precaution warn that novelty comes with uncertainty and ignorance about societal effects. The promoters then often react by downplaying the novelty. For instance, they argue that genetic engineering is really nothing new; nature itself has been doing it through the millennia and we're just doing it faster. Next, technology critics refocus on problems in the existing situation. And so the moves continue. The steps in this pattern have become expected as

---

[16] Previous versions of this diagnosis can be found in Van der Weele, 2008 and Van der Weele & Keulartz, 2009

[17] Swierstra & Rip, 2007

moves in a game, the authors write, and the pattern creates the positions of proponent and opponent, in which every move will be interpreted as a pro- or counter move. The innocent bystander who inquires into side effects is immediately labelled as an opponent.[18] In short, the role division has resulted in a rigid pattern of argumentation in which ethics has got the role and the reputation to put a brake on technology, though in fact many of the arguments in favour of technologies are just as ethical.

Naturally, ethics is not happy with a small and secondary role; it seeks to enlarge its influence. One important line of approach has been to argue that ethics should not consider the contents of science as a black box that can only be accepted as given. Instead, ethics should be involved in a much earlier phase, more "upstream" in the process, when technologies are still in the making, so that moral considerations can inform the design of the technology, including its social embedding.[19] Upstream participation fits in with values of democratization. A characteristic worry is that the voice of lay people is insufficiently heard and that, as a consequence, their life practices are invaded by new technology in uninvited and distortive ways. Scientists should talk with others, and involve citizens in the introduction of technological innovations.[20] An important implementation of this line of thought is the idea of participatory technology design: a variety of stakeholders is involved in the development of a specific technological development, say genetic screening or diagnosis.

In theory, a democratic engagement with science could be quite radical and involve choices about the themes that science should address.[21] In practice, however, democratic engagement with science and technology typically takes the form of deliberation about the implementation of concrete technologies and thus does not represent a real departure from the science first-model.

---

[18] Swierstra & Rip, 2007, p. 9

[19] A characteristic articulation can be found in Wilsdon et al. 2005

[20] Wilsdon et al. 2005, p. 23

[21] Arguments for a thoroughly democratized science can be found in Kitcher, 2001.

Starting from the world

Bioethics can also primarily be defined in relation to the world. Bioethics thus approached is that branch of ethics that deals with the "bio"- aspects of the world, the bodily aspects of living organisms: life and death and illness, growth, food, diversity, ecology. Important bioethical issues easily come to mind: global and local injustices in food and health opportunities, the fate of animals in intensive husbandry, the deterioration of environments. In the present world, an analysis of moral priorities typically takes a global perspective: through global exchange and information, the huge inequalities in the world, as well as the global problems of biodiversity, sustainability and climate have become conspicuous as bewildering moral issues. This understanding of bioethics informs many initiatives. Examples include the journal Development World Bioethics, dedicated to developing countries' bioethical issues, and the Eubios Ethics Institute, aimed at the cross-cultural development of values aimed at the good life.[22] When such a bioethics takes a pragmatic approach, it will be problem-oriented and it will look for solutions wherever they can be found. When it turns to science and technology, it will ask how they can help when people suffer, when animals suffer, when ecosystems suffer.

But when we wonder what science does to overcome global inequalities, the problems intensify. A well known way to characterize these problems is in terms of the "10/90 gap", which was originally formulated for health research: 90 percent of global resources in health research are spent on health benefits for 10 percent of the world population. The richest 10 percent, that is. The term was coined in the 1990s by the Global Forum for health research, which presented the figure as a reasonably precise estimate. Since then, the landscape of health research has changed in important ways, for example through the activities of the Bill and Melinda Gates Foundation. Yet the Global Forum notes on its website that the health needs of poor countries are still "grossly underresourced". The term 10/90 gap no longer represents a quantitative estimate, but has

---

[22] For an example see Macer, 1998, who defines bioethics as love of life.

become "a symbol of the continuing mismatch between needs and investments" in health research and elsewhere.
An ethics that starts from problems of the world has a sharp eye for these gaps, in science as well as in ethics itself. With regard to ethics, the approach implies criticism on the ELSI research agenda because it follows science and therefore predominantly addresses the concerns of rich countries, reinforcing the 10/90 gap in science. Simple counting reveals the gap. Edward Woodhouse and colleagues conclude that "during the past ten years only three of 366 published articles in Social Studies of Science and in Science, Technology & Human Values have dealt with agriculture in Low Development Countries."[23] Volkert Beekman and I have made similar observations. On the basis of counting abstracts in three conferences of the European Society of Agricultural and Environmental ethics, we noted an extreme neglect of global issues.[24] Again similarly: if the journal Bioethics had not demonstrated a 10/90 gap, there would have been no need for Development World Bioethics.

Turning from ethics to science: James Flory and Philip Kitcher have argued that the 10/90 gap shows that science is not functioning properly.[25] The goal of science is not simply to discover truth, they say, because there is far too much truth; the goal is to find significant truth. The two sources of significance, as they see it, are curiosity and practical problems. And when we think of significant practical problems, it would be a great moral error to confine ourselves to the affluent world. The direction of science should not be guided by the historical accident of where the resources for scientific research are located. The proper scope for science is global, and the 10/90 gap is a sign that science is not efficiently directed toward the most significant questions. Not all scientists can do something about this, but in health research the possibilities are abundant. For example, genomics can contribute to the search for vaccines for tuberculosis and malaria. More generally, life scientists do have many options to make science more significant. Being scientists, they have the obligation to do so, and so have the institutions that set research agendas. Just as, on a higher level, legislators have the responsibility

---

23 Woodhouse et al. 2002, p. 303
24 Beekman & Van der Weele, 2004
25 Flory and Kitcher, 2004; for background argumentation see also Kitcher, 2001

to modify the conditions under which pharmaceutical companies and research funds allocate their research.

Others too have started from the 10/90 gap in order to determine the goals of genomics. Peter Singer and Abdallah Daar want to "harness" genomics and biotechnology to improve global health equity, as the title of their agenda-setting paper aptly summarizes.[26] They call upon the G8 leaders to support implementation of an African initiative to prevent a "health genomics divide" from developing, and they mention Cuba as a country that illustrates the possibilities: it has developed a meningitis B vaccine and biotechnology ranks third, behind only sugar and tourism, among Cuban industries. The call is clear: more effort is needed to make genomics work for global health equity. For that purpose, Singer and Daar proposed to erect a Global Genomics Initiative (GGI), to raise money and coordinate efforts.

Bioethics thus understood takes the lead in science-ethics relations and acts as a guide for scientific agendas. Yet such proposals immediately meet with a gap between moral demand and scientific supply. One problem is that solving moral problems does not necessarily require results from the frontiers of science, or genomics more specifically; other roads may be more practicable and more plausible. As Adele Langlois notes in a paper on the governance of genomics,[27] the emphasis in Singer and Daar's GGI-proposal on the crucial role of genomics for health care is open to the criticism that there are more immediately effective remedies: clean water, adequate nutrition, sanitation, etc. This problem is not restricted to the GGI. Though an investigative, experimental, innovative and creative approach is highly needed to address global problems, the questions are very different from those at the frontiers of science. The Ghanaian-Dutch program of Health Research is an example of a project that suffered from the consequences of this gap. The unique characteristic of this program of development assistance was that it was a completely demand driven research program: research priorities were established, in interactive ways, in Ghana. The idea was to establish North-South research partnerships in which all the relevant

[26] Singer & Daar, 2001

[27] Langlois, 2006. p. 54; Van der Weele, 2009-2010, post 28

parties were present. But a review of this program made it clear that it did not work out this way.[28] Right from the beginning, Dutch health researchers expressed difficulty to translate the demands from Ghana into challenging research questions in their fields of research. In the ensuing "clash of perspectives", most of these researchers left the program. In their place, management-oriented development researchers came on board, who were able to focus on health systems rather than disease treatment. They enabled "capacity building" in research in Ghana and from practical points of view some of these efforts were certainly successful. But they hardly involved medical science. The reviewers conclude that institutional rigidities forced the Dutch researchers to adhere to strict scientific and publication objectives and prevented their support to more innovative research approaches.

The example illustrates that the gap partly results from the incentive system in science. Science at its frontiers has a very forceful dynamics of its own, which is fuelled, among other things, by a strong imperative for scientists to publish in highly ranked journals or perish. The imperative strengthens gaps with the needs of the world, because it is a forceful goal in its own right. The Dutch malaria researcher Bart Knols has estimated that in his field, malaria research, only about 5 percent of the scientific papers are really useful in the fight against malaria, due to the fact that the number one task in scientific research is not to solve problems but to publish at the frontiers of scientific knowledge.[29]

This incentive system, in which specialized work at the frontier is held in much higher esteem than interdisciplinary and integrative research, is closely associated with the assumption that the frontiers of science are the main source of innovation.[30] From the point of view of the 10/90 gap this is a problematic assumption, which overlooks the enormous innovative potentialities of combining existing knowledge in new ways or solving problems through low tech creativity. The System of Rice Intensification (SRI) is an example. It is a low tech innovative method of rice cultivation, in fact a methodology rather than a technology, developed by a French

---

[28] Engel and Keijzer, 2006; see also Van der Weele, 2009-2010, post 8.

[29] Quoted in Van der Weele, 2009-2010, post 8

[30] For a description and criticism of this assumption, see Soete, 2008

Jesuit in Madagascar in the 1980s. It involves the careful planting of very young rice plants at wider spacing than usual, in soil that is moist but not wet. The system leads to great increases in yield and to reductions in the amounts of seed and water that are needed. Improved soil health is considered to be the key of the success of SRI. Responses from the side of agricultural science were mainly skeptical. But through the efforts of Cornell biologist Norman Uphoff in the 1990s, the system began to draw attention and from about 2000 onward it is being applied in an ever increasing number of countries.[31]

Yet Norman Uphoff has argued that as a result of the preoccupation with "genocentric strategies", many opportunities in agro-ecology are in fact being overlooked.[32] Gaëtan Vanloqueren and Philippe Baret offer the same observation, arguing that incentives in agricultural research encourage genetic strategies, while "locking out" agro-ecological innovations.[33] Publication incentives are again part of the problem. Vanloqueren and Baret notice a great difference of academic prestige between the two fields, which is reflected in the impact factors of their journals. The journals in which agro-ecological research is published do not score higher than 4.5, while genetic engineering is in journals with impact factors "as high as 29.3".[34] They see the high academic value of specialization as an important underlying reason; genetic research fits in perfectly with it, while agro-ecological research is often interdisciplinary and therefore held in lower academic esteem.

From priority dispute to cooperative inquiry

So far, I discussed two sources for bioethical agendas. The first is science. The resulting moral agenda of bioethics is an ELSI agenda and deals with the implications of developments in the life sciences. In the second, the moral agenda of bioethics stems from an analysis of the world. In the way I presented it, the priority in the second

---

[31] Uphoff, 2007
[32] Uphoff, 2007, pp.231-33
[33] Vanloqueren & Baret, 2009
[34] Vanloqueren & Baret, 2009, p. 979

model looks simply reversed: ethics now comes first, and turns to science for help… and meets with huge obstacles.

Obstacles that block solutions to the 10/90 gap certainly deserve much attention, but so does their framing. It is not inevitable to frame them through priority disputes between science and ethics; neither is such a priority dispute the inevitable outcome of the distinctions in this paper. Science and ethics can also be seen as cooperative partners in a joint search for a better world, while alternating in taking the lead.

Thoughts about the spirit of such a cooperative attitude are abundantly available in the work of John Dewey.[35] His thinking gives primacy to daily life, or "primary experience", which he sees as being guided in large part by habit and routine. Problems and doubts are the starting points of "secondary experience", the space of reflection and inquiry. For an example, think again of the Dutch living quietly along in their deep polders, without a thought about the North Sea above their heads; a life of primary experience. Meanwhile, the Dutch history of large water problems has led to the existence of Rijkswaterstaat, a large organization of secondary experience in the continuous search for new solutions to water problems.

Secondary experience gives rise to inquiry, which turns problems and doubts into questions that can be investigated in systematic ways, but which in Dewey's view should always remain an activity in the service of daily life. Science is not the only form inquiry takes; ethics too should be understood as inquiry. It is not a set of fixed values or principles, but an activity that searches for what is good. In this search, concepts and principles are habits, or tools, for thinking, always open to improvement from the perspective of their use for understanding and improving the world. There is thus no unambiguous boundary between science and ethics. There are differences, of course, but both are connected with daily life in basically the same way, through the aim of problem solving and the spirit of inquiry.[36] Dewey proposes to use every road that is available in order to make the world a better place.

In such a spirit of cooperative inquiry, primary experience is the place of origin of the agendas in science as well as ethics, but there is

---

[35] For overviews of his thought see Boisvert, 1998 and Campbell, 1995

[36] Campbell, 1995, p. 111

no fixed thought of who should follow who. For example, ethics and science are not connected as ends and means; science creates values and ends as well as means, and philosophy also creates means, in the form of tools for thinking. Dewey is critical of a deep split between ends and means anyway. Not only do ends and means constantly change, interpenetrate and overlap, but also are they equally important from a moral point of view. Think of war as a means to establish democracy. There are no absolute ends and no unimportant means and both must be open to evaluation and reconsideration in a scientific spirit.

Priority is not a matter of principle in this climate. Different groups and people can take turns as well as cooperate in taking initiatives. Current examples of such cooperation between heterogeneous partners vary widely, from the work of the Gates foundation, public-private-partnerships in the fight against neglected diseases,[37] to science-ethics collaboration in the use of tissue engineering to develop alternatives for animal testing and to develop *in vitro* meat.

Early steps in research on *in vitro* meat may serve to illustrate briefly how cooperative relations between ethics and science may develop in interaction. Given the increasing moral and practical problems associated with "normal" meat production, some people started to think of tissue engineering as a potential solution: it might be possible to grow muscle tissue (meat) directly from animal stem cells. Research on *in vitro* meat, on a very modest scale, started in response to such morally motivated demands from society. Ethicists joined the call. Patrick Hopkins and Austin Dacey were the first to review moral arguments for and against *in vitro* meat.[38] They argue strongly in favour of its development; they find it not just interesting, but something that we may be morally required to support. They recognize the relevance for science-ethics relations, noting that 'morality is not something that must simply respond to new technologies as they arrive, throwing us into confusion, but rather that morality may champion and assist in the development of new technologies.' This remark again illustrates that the division of tasks that we had gotten so used to, in which science frontiers

[37] Moran et al. 2005

[38] Hopkins and Dacey, 2008

optimistically set the agenda and society and ethics respond with critical scrutiny, is increasingly diagnosed as being unsatisfactory. The alternative I discussed is that ethics sets an agenda of its own and then addresses science. This need not result in a struggle between science and ethics. The strengthening of cooperative interactions between morality and technology is a source of moral hope and inspiration in its own right.

Acknowledgements

Work on this chapter has been conducted as part of the research project "Towards a *lingua democratica* for the public debate on genomics" in which I participated. It was subsidized by the Netherlands Organization for Scientific Research (NWO), project no. 050-32-570. The chapter also contains material from the project "How to overcome the 10/90 gap in genomics research", subsidized by the Centre for Society and Genomics. The paper builds on exchanges with many people. I especially thank project leaders and discussion partners of the two projects just mentioned: Peter Derkx and Harry Kunneman of the *lingua democratica* project, Michiel Korthals and Henk van den Belt of the 10/90 project.

Literature

Avard, D, L. M. Bucci, B. Coly & A. Saginur (2006). Review of Research in Canada and the United States and Synthesis of Key Informants Views. Montreal: ERA-SAGE Research Area on Ethical, Legal, and Social Aspects of Human Genomics.

Beekman, Volkert & Cor van der Weele (2004). *Naar een Gereedschapskist voor Constructieve Ethiek.* Den Haag: LEI.

Boisvert, Raimond D. (1998). *John Dewey: Rethinking our Time.* Albany: State University of New York Press.

Campbell, James (1995). *Understanding John Dewey. Nature and Cooperative Intelligence.* Chicago: Open Court.

Collins, Francis S., Eric D. Green, Alan E. Gutmacher & Mark S. Guyer (2003). "A Vision for the Future of Genomics Research: A Blueprint for the Genomic Era," *Nature* 422: 835–847.

Dreger, Alice (2000). "Metaphors of Morality in the Human Genome project." In *Controlling Our Destinies: Historical, Philosophical, Ethical, and Theological Perspectives on the Human Genome Project.* Edited by P. Sloan. Notre Dame: University of Notre Dame Press: 155-184.

Engel, P. and N. Keijzer (2006). *Research partnerships: who decides? Review of a design process.* The Hague. Available at http://www.ecdpm.org/ Web_ECDPM/Web/Content/Navigation.nsf/index2?readform&http://www.ecdpm.org/Web_ECDPM/Web/Content/Content.nsf/19caebbe8dae7428c1256c7900531415/ee40ad6e8ea3c0c7c12572c00034f6bd?OpenDocument

Flory, James H. & Philip Kitcher (2004). "Global health and the scientific research agenda." *Philosophy and Public Affairs* 32, no 1: 36-65.

Hopkins, Patrick D. & Austin Dacey (2008). "Vegetarian meat: could technology save animals and satisfy meat eaters?" *Journal of Agricultural and Environmental Ethics* 21: 579-596.

Jeffrey, Julie R. (1998) *Frontier Women: Civilizing the West?* New York: Hill and Wang.

Kitcher, Philip (2001). *Science, truth and democracy*. Oxford: Oxford University Press.

Langlois, Adele (2006). "The governance of genomic information: will it come of age?" *Genomics, Society and Policy* 2, no 3: 49-63.

Macer, Darryl (1998). *Bioethics is love of life. An alternative textbook*. Christchurch, N.Z : Eubios Ethics Institute.

Moran, Mary, Anne-Laura Ropars, Javier Guzman, José Diaz & Christopher Garrison (2005). *The New Landscape of Neglected-Disease Drug Development.* London: Wellcome Trust.

Singer, Peter & Abdallah Daar (2001). "Harnessing Genomics and Biotechnology to Improve Global Health Equity." *Science* 294: 87–89.

Slotkin, R. (1992) *Gunfighter Nation. The Myth of the Frontier in Twentieth-Century America.* New York: Atheneum.

Soete, Luc (2008). "International Research Partnerships on the Move." In *Knowledge on the Move. Emerging Agendas for Development-oriented Research.* Edited by H. Molenaar, L. Box, R. Engelhard. Leiden: International Development Publications: 33-48.

Swierstra, Tsjalling and Arie Rip (2007). "Nano-ethics as NEST-ethics: Patterns of moral argumentation about New and Emerging Science and Technology." *Nanoethics* 1: 3-20.

Turner, F.J. (1893) The Frontier in American History. Chapter 1. Available at: http://xroads.virginia.edu/~HYPER/TURNER/

Uphoff, Norman (2007). "Agroecological alternatives: Capitalising on existing genetic potentials." *Journal of Developmental Studies* 43, no. 1: 218-236.

Vanloqueren, Gaëtan & Phillippe Baret (2009). "How agricultural research systems shape a technological regime that develops genetic engineering but locks out agro-ecological innovations." *Research Policy* 38, no.6: 971-983.

Van der Weele, Cor (2004). *Hoe vreemd is de Beemster. Vanzelfsprekendheden en de toekomst van water in Nederland.* Den Haag: LEI.

Van der Weele, Cor (2008). "Moral agendas for genomics: how to find the blind spots?" *Configurations* 16, no.1: 117-135.

Van der Weele, Cor (2010). "The 10/90 gap: what can we do?" Researchblog 2009-2010, available at http://tenninetygap.wordpress.com/

Van der Weele, Cor (2010). "In vitro meat: promises and responses. Cooperation between science, social research and ethics" In *Global food security: ethical and legal challenges.* Edited by C.M. Romeo Casabona, L.E. San Epifanio & A.E. Cirión. Wageningen: Wageningen Academic Publishers.

Van der Weele, Cor & Jozef Keulartz (2009). "Heroes of agricultural innovation." *Genomics, Society and Policy* 5, no. 3: 59-73.

Ward, Diana Raines (2002). *Water wars: Drought, flood, folly and the politics of thirst.* New York: Penguin Putnam.

Wilsdon, James, Brian Wynne & Jack Stilgoe (2005). *The Public Value of Science; or How to Ensure That Science Really Matters.* London: Demos.

Woodhouse, Edward, David Hess, Steve Breyman & Brian Martin (2002). "Science Studies and Activism: Possibilities and Problems for Reconstructivist Agendas." *Social Studies of Science* 32, no.2: 297–319.

Wouters, Arno (2005). *ELSA Onderzoek in Nederland.* Nijmegen: Centre for Society and Genomics.

Zwart, Hub (2009). "The adoration of a map: reflections on a genome metaphor." *Genomics, Society and Policy* 5, no 3: 29-43.

6

The Lack of Access to Medicines and the Health Impact Fund scheme

Andries De Smet, Sigrid Sterckx and Julian Cockbain

> The task is huge. Take Africa. It has 34 of the 50 poorest countries in the world and suffers 24% of the global disease burden. It has just 3% of the world's health workers. [...]
>
> Two thirds of all people living with HIV live in Africa. In Africa, a child dies of malaria every 30 seconds. (Witty 2009: 1)

Introduction

In August 2008, Thomas Pogge, Professor of Philosophy and International Affairs at Yale and leading thinker on global ethics and human rights issues, and Aidan Hollis, Associate Professor of Economics at the University of Calgary, launched a proposal for a Health Impact Fund (HIF), a scheme to facilitate the introduction of new medicines aimed primarily at diseases of the developing world. (Hollis & Pogge 2008)

This chapter aims to provide some background to the access to medicines crisis, to give a brief overview of some suggestions for promoting access to medicines in poor countries, to explain the HIF mechanism and to discuss some of its perceived advantages and shortcomings vis-à-vis alternatives.

Although we conclude that many hurdles have to be overcome and that the HIF scheme needs improvement, we strongly support any attempt to improve healthcare in the developing, and indeed the developed, world. The HIF scheme contains several important elements towards a solution.

The access to medicines crisis

The developing world faces several problems in relation to disease (Pécoul et al. 1999): there may be no medicines effective against the disease because of a lack of R&D; many of the existing medicines are toxic and/or out of date and/or not adapted to the circumstances on the ground; problems of resistance are widespread; the available medicines may be too expensive for the patients or the country to afford; there may be no efficient system for ensuring that the appropriate medicines reach the patients; and lack of education, prejudice, hunger or insanitary conditions may cause disease to be more serious or prevalent than would be the case in an industrialised country. Of these different problems, perhaps only the last is beyond the ability of pharmaceutical companies (hereinafter 'Pharma') to address. None are beyond the ability of national governments, perhaps with the assistance of supranational governmental institutions, to address.

Within this context, in 2000 the United Nations launched the Millennium Development Goals. These goals cover a wide range of topics, varying from the eradication of extreme poverty and hunger to empowering women. Goal 8 focuses on developing a global partnership for development. Goal 8E, one of the constituting targets, specifically addresses the access to medicines crisis: "In cooperation with pharmaceutical companies, provide access to affordable essential drugs in developing countries".[1] The target date for these goals is 2015. How close to reaching this target are we?

Between 2007 and 2011 a number of surveys was conducted, using the World Health Organisation/Health Action International standard methodology.[2] The data imply that generic medicines are

---

[1] See http://www.un.org/millenniumgoals/global.shtml

[2] See http://www.haiweb.org/medicineprices

only available in 51.8 % of the public sector facilities of developing countries. Availability in the private sector is higher, more specifically in 68.5 % of facilities. However, availability is not the whole story. The surveys indicate that the medicines are sold at a much higher price than the international reference price (IRP). On average, people have to pay 2.6 times the IRP in the public sector, as compared to an average of 5 times the IRP in the private sector[3]. Needless to say, most patients in the developing world simply cannot afford this.

Next to issues of availability and affordability, the access to essential medicines crisis is caused by the inadequate incentive structure of the pharmaceutical industry. Because this industry is profit-driven, the focus of the R&D will be on the most lucrative remedies, rather than on those with the largest health impact. Poor countries are smitten with infectious and parasitic diseases, while in rich countries these diseases represent but a small fraction of the disease burden (measured in so-called *Disability-Adjusted Life Years* or DALYs). Given the current structure, a mismatch occurs between the needs of the poor and the investments of the pharmaceutical companies, the so-called "10/90 gap". With this concept the Global Forum questioned the fact that less than 10% of the total health-related R&D is devoted to the major health problems representing 90% of the global disease burden.[4] It has become widely accepted in theoretical analyses of the problem, but those who suffer from the inflictions in question see little change.

The following data exemplify this situation. Between 1975 and 2004, 1,556 new chemical entities were introduced to the market globally. Only 20 of these – i.e. a mere 1.3 per cent – were for tropical diseases and tuberculosis, even though these represent 12 per cent of the total disease burden (Chirac & Torreele 2006). In more recent years (2000-2011), some progress has been made, but according to the Drugs for Neglected Diseases *initiative* (DND*i*) and Médecins Sans Frontières (MSF) only 3.8 % of newly approved drugs (excluding vaccines) were for tropical diseases, TB, and other neglected infections, which together account for 10.5 % of the global disease burden. They also point out that much of the progress is a

---

[3] See http://www.who.int/medicines/mdg/mdg8report2012_en.pdf

[4] See http://www.globalforumhealth.org

result of drug reformulations and repurposing of existing drugs. When looking at the newly developed medicines, they conclude that only four of the 336 new chemical entities developed between 2000 and 2011 were for the treatment of neglected diseases. Still more recent, as of December 2011, only 1.4 % of a total of nearly 150,000 registered clinical trials were focused on neglected diseases.[5]

Even though the admirable and highly professional policy advocacy work of Non-Governmental Organisations like *Médecins Sans Frontières* (MSF) has resulted in important advances, radical changes are required to the current approach to incentivising pharmaceutical R&D, or this outrageous and absolutely unjustifiable situation will only deteriorate.

Pharmaceutical industry R&D and patenting

By Pharma's reckoning, to find and bring to market a new medicine costs in the region of $1 billion. From the start of research until market launch may take ten years or so. Thus for example, in a policy statement of November 2007, GlaxoSmithKline, one of the world's most important pharmaceutical companies, commented:

> The cost, time and risk in bringing a product to market is huge. For every 10,000 compounds that are tested for pharmaceutical activity, only 3 reach the market and then only one in every 3 drugs which reach the market is profitable. It costs on average around $1.2 billion to bring a drug to market and typically takes 10-12 years. (GSK 2007: 2)

Even though these figures are heavily criticised for being highly exaggerated (Public Citizen 2003; Relman & Angell 2002), the risks and costs of drug development are indeed very significant. To make huge investments over a long period of time requires government financing (think of Kennedy and the first moon landing), a Bill Gates, or a company which can see a reasonable chance of recouping the investment, generally through sales at a considerable profit facilitated by patent protection.

---

[5] See http://www.dndi.org/media-centre/press-releases/354-media-centre/press-releases/1515-millions-patients-against-neglected-diseases.html

While the US, the EU and Japan together represent only 15% of the world population, they make up 87% of the global pharmaceutical market, which was forecast to be over $735 billion in 2008 (IMS 2008). The corporate officers of public companies have a fiduciary duty to their shareholders, i.e. a legal obligation to optimise the value of the shareholders' investment. As a result, Pharma tends to invest in R&D for medicines which are effective against diseases or ailments suffered by a large number of patients in industrialized countries who, directly or through health insurance or national health systems, can afford to pay prices far beyond the capacity of the average citizen of developing world countries.[6] Accordingly, R&D of medicines to treat diseases affecting predominantly or only the developing world is severely neglected.

Lack of access to medicines, not only in poor countries but also in many industrialized countries, is also linked with the operation of the patent system (Sterckx 2007).To quote again from the recent speech of GlaxoSmithKline's CEO at Harvard Medical School:

> Our industry has treated IP [intellectual property, in this context patent law] as if it is written in stone. Yes it is vital but we shouldn't forget that IP is always a balance, a grand bargain, with society. (Witty 2009: 2)

Indeed, in the past, the patent laws of many countries did not permit drug substances as such to be patentable, only processes for preparing them. To "design around" such process patents, i.e. to invent alternative, non-infringing processes, was normally relatively straightforward, and so the pharmaceutical companies in such process-patent countries could produce cheaper, generic, copies of new medicines for sale in their own countries, in other 'only-process-patent' countries, and in countries where the originator company had failed to seek patent protection. A large proportion of the medicines sold in Africa were produced in this way by Indian companies for example.

However, in the *Marrakesh Agreement establishing the World Trade Organization* (1994), an Annex was introduced –essentially under

---

[6] Bear in mind, for example, that the annual treatment cost with Herceptin ®, a medicine for the treatment of certain breast cancers, is about $30,000.

pressure from the United States, itself under pressure from companies such as Pfizer– which requires all countries which would be party to the World Trade Organization to permit patenting of drug substances as such. This Annex, the Agreement on Trade Related Aspects of Intellectual Property, better known as the TRIPs Agreement (WTO 1994), means that in future the ability of generic pharmaceutical companies to sell affordable copies of new medicines will be severely restricted.[7]

Earlier proposals for addressing the lack of access to medicines

Several suggestions for addressing the lack of access to medicines have been made in the recent past and some are already in place. All are inadequate in one or more respects.

The most obvious solution for drug discovery and development is for it to be funded by *the government of the country whose citizens suffer from the disease in question.* For the developing world however, this is a pipedream as such governments generally have more pressing demands, legitimate or otherwise, on the money they may have.

To some commentators, it is equally obvious that *'big Pharma'*, in view of the 'obscene profits' it makes, should carry out such R&D as part of its obligation to the community. Again, however, this is a pipedream in view of the fiduciary duty of management to shareholders. To change this legal obligation would be possible, but would require international synchronisation and strict enforcement to avoid companies in one country being disadvantaged in competition against companies from others. Although desirable and attractive, this solution seems unlikely in the visible future.

Perhaps such R&D should be seen as a role for *academia and/or public research institutes*? After all, much if not most of the fundamental research underlying the pharmaceutical industry has been and is being done in this context (NIH 2000). However, academic research needs to be funded, and such funding comes from national governments, supranational institutions, industry and charities. Charity funding of R&D of medicines for neglected diseases is uncertain given the sums

---

[7] With regard to the developments leading up to the TRIPs Agreement: see Sell (1998), Matthews (2002) and Drahos & Braithwaite (2002).

at issue,[8] industrial funding is unlikely for the abovementioned reasons, and state funding of this kind of R&D is either unlikely or is subject to the politicians' need to justify their actions to their electorate when facing re-election, i.e. uncertain. Moreover, even if basic research in academia to identify potential medicines is government financed, academia does not have the resources or skills, or even desire perhaps, to take a project through development, clinical trials and manufacturing. Thus, while academia and governments have a role to play, this alone is not enough to remedy the lack of access to medicines faced by hundreds of millions of people.

Currently proposed solutions

The solutions proposed until now essentially involve: (a) patent protection; (b) compulsory licensing of patents; (c) voluntary licensing of patents (including patent pools); (d) parallel inducement; (e) orphan drug protection; (f) supranational funding of R&D; (g) advance market commitments; and (h) prizes. We will briefly comment on each of these schemes.

Option (a) – Patents

This represents the status quo – the patent system has worked to deliver medicines for diseases and ailments of the affluent and the military, but it is clearly inadequate to meet the needs of patients in the developing world or indeed the needs of those suffering from rare diseases anywhere in the world.

Option (b) – Compulsory licensing of patents

Before the entry into force of the WTO-TRIPs Agreement, many countries made it possible for licences to be granted to produce patented medicines even when the patentee was unwilling to grant

---

[8] While the work of the Bill and Melinda Gates Foundation is to be strongly applauded, we cannot rely on such support to continue at the level required to address global health problems in full – a more structural solution is required.

licences voluntarily. One of the main developments instituted by TRIPs was a severe restriction of this ability. While some relaxation occurred with the so-called Doha Declaration on *The TRIPs Agreement and Public Health* (2001),[9] compulsory licensing is still relatively unused and does not address the issue of encouraging R&D in new drugs.

Option (c) – Voluntary licensing of patents (including patent pools)

Patent holders may of course voluntarily license others under their patents. This may be attractive if the patentee does not wish itself to commit to R&D against a particular disease, while still wishing to keep the option open to benefiting from R&D success by the company that does, e.g. benefiting through royalties, access to cross-licences, access to compound libraries, etc. One way to achieve this is to create a patent "pool", where a group of patentees commit to allow others access to their patents under conditions they are comfortable with. This is the scheme recently proposed by Andrew Witty, CEO of one of the world's foremost pharmaceutical companies, GlaxoSmithKline. Witty's proposal, although a welcome development, was restricted to a shortlist of specified diseases, excluding HIV, and Witty notably stated that "[i]f – as we hope – something new comes out of such research the full benefits should go *solely* to the [least developed countries]. The terms and spirit of the pool should ensure this happens" (Witty 2009: 2, emphasis added). Unfortunately, this does not imply that benefits would be reaped by the poor outside the least developed countries.

For these reasons, the healthcare NGO *Médecins sans frontières* has rightly criticised GSK's offer as inadequate to deal with the problem (MSF 2009).

Option (d) – Parallel inducement (including fast track review)

One of the more egregious proposals for improving access to medicines for the poor in the least developed countries, involves the reward of the originator company with an otherwise unrelated inducement. One suggestion has been priority review vouchers (Ridley et al. 2006) whereby the company obtaining market approval for a medicine against a neglected disease would receive entitlement

[9] WTO Document WT/MIN(01)/DEC/1, 20 November 2001.

to priority in US regulatory approval review for another medicine, e.g. one aimed at developed world diseases or ailments. This clearly gives the originator an unfair advantage in the developed world market. Another such proposal has been made by Professor John Barton of Stanford University and Jeff Kindler, CEO of Pfizer, in their letter of 13 August 2008 to the US Senate Finance Committee (Kindler and Barton 2008). One key to the Barton/Kindler proposal is that: "Developed-world nations would commit themselves to develop detailed mechanisms to ensure that their government pharmaceutical purchasing authorities pay an adequate price to encourage research" (Kindler and Barton 2008: 1). Translated, what Pfizer is asking for is that one inducement to carry out R&D for neglected diseases would be for developed world governments to refrain from using their purchasing power to negotiate downwards the prices Pharma wishes to charge in the developed world for *all* its products.

Option (e) – Orphan drug protection

This is also the status quo in the United States. It works to deliver medicines to treat rare diseases of the wealthy or those with health insurance, but it is not a solution to the needs of the developing world.

Option (f) – Supranational funding of R&D of medicines for neglected diseases

This proposal is exemplified by the Medical Research and Development Treaty (Love 2005). However, it relies on supranational commitment to support R&D as well as on modifications to patent laws and freedom to patent. Moreover, as Hollis and Pogge point out, it seems to give flexibility in funding allocation to an extent that would "enable governments to make resource allocations based on domestic political interests, rather than global health needs" (Hollis and Pogge 2008: 104).

Option (g) – Advance market commitments

This is the subject of the proposal that developing world governments should commit to ordering X million doses at a dose price of $Y of a medicine to treat disease Z when the medicine is market-approved (Kremer and Glennerster 2004). This scheme faces three particular problems: the uncertainty as to whether the government will meet its obligation when the time arises; the risk that a competitor wins market approval first; and the uncertainty as to whether medicines delivered to the country will actually reach the intended patients if the local healthcare system is inadequate. Some may argue that the second problem, the risk of losing the race, is also faced by Pharma in normal practice – i.e. under option (a). However, a 'me too' drug may still sell well, especially if competitively priced, hence in normal practice Pharma does not face total loss by coming second to the developed world market. Under scheme (g), though, the first company to arrive on the market may capture all the available advance market commitments.

Option (h) – Prizes

The final option involves a system whereby Pharma is rewarded by a monetary prize for bringing a new medicine to market rather than (only) by the grant of a patent monopoly. One such system, which replaces patents, has been proposed by James Love, Director of the NGO *Knowledge Ecology International*,[10] and is currently before the US legislators as the Medical Innovation Prize Act of 2007 (Sanders 2007). However, being entirely non-compliant with the abovementioned WTO-TRIPS Agreement (1994), this scheme has a snowball in hell's chance of seeing daylight.

The prize is a tried and tested system, which was used for example with the determination of longitude in the 18th century.[11] With regard to R&D of medicines for neglected diseases, if the prize would be awarded by a national government or charity, its availability on delivery would be doubtful in view of the uncertainty as to whether the government will meet its commitment when the time arises (again, as demonstrated by the case of longitude). Justifiable doubt would provide the management of Pharma with further bias, if more

---

[10] KEI is an organization committed, inter alia, to improving the availability of medicines and particularly to schemes involving supranational funding of medicine development or payment by prize rewards rather than by the grant of (patent) monopolies. See www.keionline.org.

[11] In the 18th century, the British government offered a prize of £ 20,000 (about US$ 4.5 M in today's currency) for an accurate means of determining longitude. The problem was solved by a clockmaker, John Harrison, working between 1730 and 1772. See Sobel (1995).

is necessary, against taking the decision to commit to R&D in relation to developing world diseases.

A new idea: The Health Impact Fund (HIF)

The HIF, as recently proposed by Thomas Pogge and Aidan Hollis, is essentially a supranationally-funded prize system intended to address the problem of affordable access to new medicines for diseases primarily affecting developing countries (Hollis & Pogge 2008).

Put very briefly, HIF involves a supranational body, majoritatively funded by developed countries, paying an annual prize for up to ten years to a company bringing a new medicine to market, with the size of the payment reflecting the reduction in global health burden attributable to the use of that medicine and with the medicine being sold at a pre-agreed and hence affordable price. As currently proposed, HIF does not preclude the company that has developed the drug or the 'originator company' from using its patents to exclude competitors, and so the originator's investment can be recovered by the combination of the profit component in the agreed price and the annual prize payments.

One key feature of the HIF is that the registrant company must commit to provide open licences to the relevant patents at the end of the reward period, i.e. within 10 years of market launch. A second key feature is the fact that no HIF reward is payable before the year after market launch. In these two features lie the two major objections to the HIF – that it does less than possible to maximise the availability of medicines at the most affordable prices and that it does little to ease the decision by Pharma to enter into an R&D project by reducing either the amount of money at risk or the period before investment is recouped if the project is successful.

The debate over HIF's potential impact on the generic pharmaceutical industry

In an *earlier form* of the scheme, HIF was intended to require the originator to *permit generic competition from the outset* and thus the originator company would have recovered its investments through smaller profits on sales and a prize which took into account the impact of *both* the originator's product and the generic copies.

In our view, introducing the pre-agreed price and permitting the use of patent monopolies represents a retrograde step in the development of HIF as the goal of HIF is to maximise reduction in global health burden, something which is facilitated by the downwards pressure on price that results from generic competition. The case of antiretroviral drugs provides a perfect example. In 2000, a year's treatment course for HIV/AIDS cost more than US$ 10,000 per person. At that stage, these so-called 'first generation' antiretrovirals were only available from the pharmaceutical companies that held the patents on these drugs. Currently, the most commonly used triple-drug antiretroviral treatment in the developing world costs less than US$ 100 per year. The reason why the price of these drugs has dropped with 99% over a period of only six years is that these products were not under patent in several countries with pharmaceutical manufacturing infrastructure – e.g. Brazil and India. This enabled local manufacturers in those countries to legally produce generic versions of the brand name medicines patented in developed countries. Moreover, those generic medicines could also be exported to other developing countries where the medicine was not covered by a patent.[12]

Pogge and Hollis's HIF scheme has attracted severe criticism from James Love of *Knowledge Ecology International.* Love, who has published his correspondence on the subject with Pogge and Hollis online (Love 2008a and Love 2008b), particularly criticises the stepping back from open licensing which occurred in the maturation of the HIF scheme.

---

[12] Unfortunately, all this is changing as a result of the entry into force of the WTO-TRIPs Agreement, which forces all member states to grant product (and not only process) patents in all fields, including pharmaceuticals. See Sterckx (2007). Moreover, in view of problems regarding rapidly spreading resistance to first-generation treatments, patients receiving such treatments increasingly need to shift to newer drugs to which they are not resistant, but which are under patent and hence hugely expensive.

Pogge and Hollis's main justifications[13] for modifying the initial form of the scheme were essentially that (a) the removal of a requirement for open licensing would make HIF more attractive to Pharma both to participate in and to refrain from challenging the scheme, and (b) HIF makes open licensing redundant as the registrant is incentivised by HIF to supply at the lowest price.

Even though the first of these arguments rings true, the second justification is clearly flawed. In this regard, it deserves repeating that generic competition has been shown in the field of anti-retrovirals to be able to bring prices down by 99% while GlaxoSmithKline in its recent offer to help access to medicines has merely offered to reduce prices in the least developed countries by "at least 75%" (Witty 2009).

James Love has expressed particular concern about the effect of HIF on the generics industry, arguing that, as structured, the downwards pressure on price encountered where generic competition arises, would be reduced under HIF:

> [T]he … HIF approach … would marginalize the generics industry in developing countries, make it harder to credibly threaten a compulsory license, raise the costs of generic suppliers (fewer economies of scale or scope), and make the political environment for [compulsory licensing] or UNITAID type [voluntary licensing] quite difficult. …
>
> Finally, it is only a minor virtue that the HIF would ask for an open license after ten years. For many products, the effective patent and product life is not much longer than this anyway, particularly given late product registrations in developing countries. (Love 2008a)

---

13 Besides the two justifications mentioned, the following were also given: (c) the entry of generic competition would make it difficult to assess the HIF reward; (d) the problem of counterfeiting of medicines would be smaller, or would be more controllable; (e) distributors would not be tempted to trade in the higher profit margin versions of the medicine rather than the lowest price to customer version; (f) the registrant would be able to recoup a higher profit in countries able to bear a higher price and so would find the scheme more attractive; (g) registrants might fear misuse of licensed technology, i.e. for the preparation of things other than the registered medicine; and (h) registrants might be constrained in their ability to sublicense by their own licensors.

Indeed, by not mandating generic competition, HIF does not give a strong enough incentive to the registrant or the scheme administrators to drive prices down.

HIF also leaves some other important problems unaddressed and raises some new ones. These are briefly discussed in the next sections, before we turn our attention to HIF's main strengths in comparison with other schemes.

Problems resulting from HIF's linkage with patents

A company which owns or has licensed all relevant patent rights to a new medicine may register that medicine under the HIF scheme and thereby become entitled to a reward under the scheme for each of the first ten years, or each remaining year of the first ten years, following first marketing authorization. Accordingly, patents must exist for the medicine or its process of preparation or use. In view of this feature of the scheme, rewards are *not payable* on products deemed to be previously known or obvious modifications of known medicines,[14] or on medicines which are or are shortly to become off-patent.

Since a company wishing to benefit from the HIF scheme must make use of the patent system, it will have every incentive to use that system to exclude competing products, whether identical or not. Moreover, since the company must have patent cover for the medicine, the decision whether to register with HIF or not will be based on the comparison of the expected profit levels should the medicine be marketed conventionally with the expected HIF rewards if the medicine is HIF registered. For a medicine with a significant expected developed world market, e.g. one effective against cancer, heart disease or age-related diseases, it seems unlikely that HIF registration would occur.

---

[14] As such products would not meet the legal patentability requirements of novelty and inventive step – see Cockbain (2007).

Problems arising from the obligations HIF registrants must accept

In return for becoming eligible for the HIF rewards by registering a medicine, the registrant must agree to sell the medicine at a pre-agreed price, which will cover production costs but will have little profit margin. A HIF registrant must also agree to license others under its patents relating to the medicine after the ten year reward period ends. Since the pre-agreed price covers production costs, the only incentive to reduce production costs is to increase sales and hence the medicine's impact on global health burden and thereby increase the reward the registrant may receive. However, once a company has begun sales of a patented medicine it has little incentive to change the production process, *unless* faced by generic competition, since regulatory approval problems may arise if the production process is changed. As we explained earlier, under HIF, generic competition is not foreseen until the ten year reward period is over.

Where there *is* generic competition, for example in countries where no relevant patents exist or where the HIF registrant has elected not to seek market authorization, the health impact of the generic products will be ascribed to the registrant in assessing the HIF reward. Where no patents exist, generic manufacturers require no licences and so fall outside the control of the registrant and/or HIF. As a result, there may be some difficulty in determining the local sales volume of the generic products and this will complicate the calculation of the component of the HIF reward which is based on the health impact resulting from sales of generics.

Pogge and Hollis also propose that HIF registrants must accept to "make a good faith effort to obtain market clearance wherever the product is needed" and to "preauthorize the HIF to seek market clearance for the product wherever the registrant has failed to do so and to subtract the cost of this effort from the registrant's next health impact reward payment" (Hollis and Pogge 2008: 14). For products useful in many countries, the registrant is thus being asked to bear the costs of obtaining market approval in all such countries, either directly or through a reduced HIF reward, costs most likely to arise in the early years of the reward period. This potentially delays the time to break-even, making the scheme less attractive.

Problems resulting from the baseline calculation of the health impact

According to Pogge and Hollis, the health impact of a HIF registered medicine would be assessed as the *attributed incremental health impact* of the medicine measured as the difference between actual health and a baseline representing the expected health level projected from the actual health at medicine launch (or two years earlier) and, in this projection, excluding the contribution of the medicine (and any other medicines sold exclusively by the registrant). For non-communicable diseases, the impact would be assessed on the basis of the treated patients, e.g. as the increase in quality-adjusted life years (QALYs). For communicable diseases, the impact would include all other indirectly affected individuals. As Pogge and Hollis admit, this would make HIF the largest health assessment agency in the world.

Unfortunately, excluding *only* the medicine itself and any of the registrant's other products from the baseline calculation makes the HIF reward susceptible to reduction in the event that *other medicines targeting the same disease* are launched during the reward period. This may cause two problems to arise. Firstly, that the decision to commit to R&D is made more difficult in view of the likelihood of the HIF reward being reduced. Secondly, the registrant is incentivised to use the patent system to exclude or delay the arrival of competing products, either 'me-toos'[15] or quite different products, or at minimum to require licence fees that counterbalance the HIF reward reduction. This would *not* achieve the goal of promoting access to medicines.

Problems regarding security of funding

As conceived, HIF would be funded by national governments committing to contribute a percentage of Gross National Income, e.g. 0.03%, over a period of at least 10 or 12 years. Before R&D is initiated based on an expectation of repayment by HIF rewards, or at

---

[15] A 'me-too' is a pharmaceutical which is closely similar, structurally, to an existing product, but which may have minor benefits, e.g. in terms of reduced side effects, enhanced uptake, etc.

least before development reaches phase III clinical trials, some certainty is needed of there being adequate sums in the HIF when the HIF rewards fall due. This involves reasonable certainty that enough countries commit and will meet their commitments and that the numbers of medicines registered will not be so high that the size of the HIF rewards will be inadequate. Payments to HIF would effectively be foreign aid and since relatively few countries are generous with foreign aid, this may appear too uncertain to "pull" drug developers into the scheme.

If the governmental commitment required is limited to 10-12 years, as it is under the current proposal, then HIF would be essentially an incitement for a 'single round' of R&D - those reaching the market first will gain a large share, those arriving later may find that the fund is drying up when they ask for their rewards over their 10 year reward period. As not all new medicines arising out of HIF would reach the market at the same time, the HIF reward funding will have to be guaranteed over a longer period.

Finally, we turn to the strong points of HIF vis-à-vis other proposed schemes.

The attraction of HIF: the reward

The HIF rewards would be paid from a fund financed by contributions from countries party to the scheme, those contributions being a fixed percentage of the contributing country's Gross National Income. The fund for each year will be divided between all HIF registered medicines currently within their ten year reward period in proportion to the effect each has had that year on reducing global health burden. The fewer the eligible medicines, the larger the reward and vice versa. Since the decision to initiate R&D on a drug is generally taken about ten years before market approval, at the time of decision the company will have no means of knowing what level of HIF reward it might receive if it commits to start work on a medicine which would not be profitable if sold conventionally.

However, R&D expense escalates dramatically as the development progresses and by the end of phase II clinical trials, the time to market approval may be short enough to make the likely size of HIF reward more estimatable. Thus it may well be the case that HIF would encourage R&D to progress to *this* stage for products that would not be profitable with conventional sales. Some research and development into diseases of the developing world does take place already[16] and HIF may stimulate the companies involved with such R&D to bring products to the market by allowing the decision to enter the most costly stage of clinical trials to be made with a reasonable expectation of sufficient reward within a tolerable period.

Moreover, that reward, coming as it would from a supranational body, could be more confidently expected to materialise then would be the case where it was dependent on the continued existence of charitable funding or on the vagaries of national politics.

The attraction of HIF: a potential solution to the "last mile" problem

A very strong point of the HIF scheme is that it takes into account what Pogge and Hollis refer to as the "last mile" problem of ensuring delivery of medicines *to the patients* (Hollis and Pogge 2008: 71-81). It achieves this by making the HIF reward proportional to the measured

---

[16] GlaxoSmithKline, for example, has a research centre dedicated to diseases of the developing world and employing 100 research scientists in Tres Cantos, Spain (Witty 2009). Likewise, in 2002 Novartis set up the Novartis Institute for Tropical Diseases (NITD) in Singapore, an operation with over 100 researchers and supporting staff (see Novartis 2008). However, such efforts must be viewed with a smidgin of cynicism. According to Witty, the Tres Cantos centre is funded in part by its partners, the Medicines for Malaria Venture and the Global Alliance for TB Medicine Development, while the Novartis web-site identifies NITD's focus as being on Dengue fever, TB and malaria. While TB and malaria are indeed problematical in the developing world, TB is also a problem for AIDS sufferers and malaria also one for developed world tourists and military - *i.e. it seems that the focus is on medicines that may find a satisfactory conventional market.* As far as Dengue fever is concerned, it may be noted that in a talk at a conference arranged by *Médecins Sans Frontières* in London in June 2005, Dr Paul Herrling, Head of Corporate Research at Novartis, speaking about NITD, admitted that NITD was concentrating on diseases that had analogues, such as West Nile and hepatitis, for which there might be a conventional market for effective medicines.

health impact – if the medicine does not reach the patients, there cannot be a health impact and hence no HIF reward.

This is undoubtedly the most original and convincing aspect of the HIF scheme.

References

Chirac P. and Torreele E. (2006), Global framework on essential health R&D, *The Lancet* 367, pp. 1560-1561.

Cockbain J. (2007), "Intellectual Property Rights and Patents", Chapter 1.26 in Taylor J. and Triggle D., *Comprehensive Medicinal Chemistry II*. Vol. 1. Oxford: Elsevier, pp. 779-815.

Drahos, P. and Braithwaite J. (2002), *Information Feudalism – Who Owns the Knowledge Economy?* London: Earthscan.

GlaxoSmithKline (2007), *Global Public Policy issues – Voluntary Licensing of ARVs*, available at: http://www.gsk.com/policies/GSK-on-voluntary-licensing-of-ARVs.pdf. [last accessed 23 February 2009]

Hollis A. and Pogge T. (2008), *The Health Impact Fund: Making new medicines accessible for all*, Incentives for Global Health.

IMS Health (2008), *IMS 2008 Global Pharmaceutical Market Forecast*, available at: http://www1.imshealth.com/web/content/0,3148,64576068_63872702_70260998_84286337,00.html [last accessed 23 February 2009].

Kindler J. and Barton J. (2008), *Letter of 13 August 2008 to The Honorable Max Baucus, Chairman of the US Senate Finance Committee*, available at http://media.pfizer.com/files/news/kindler_baucus_letter_081308.pdf [last accessed 1 March 2009].

Kremer M. and Glennerster R. (2004), *Strong medicine: creating incentives for pharmaceutical research on neglected diseases*, Princeton: Princeton University Press.

Love J. (2005), *Medical Research and Development Treaty*, available at www.cptech.org/workingdrafts/rndtreaty4.pdf [last accessed 23 February 2009].

Love J. (2008a), *Impact of the HIF on generic industry in developing countries*, available at www.keionline.org/blogs/2008/11/20/impact-of-hif-on-generics/ [last accessed 28 February 2009].

Love J. (2008b), *Why the HIF rejected open licensing*, available at www.keionline.org/blogs/2008/11/19/why-hif-rejected-open-licensing/ [last accessed 28 February 2009]

Matthews D. (2002), *Globalising Intellectual Property Rights – The TRIPs Agreement.* London: Routledge.

Médecins Sans Frontières (2009), *MSF response to GSK patent pool proposal.* Press release, 16 February, available at: http://www.msfaccess.org/media-room/press-releases/press-release-detail/?tx_ttnews%5Btt_news%5D=1532&cHash=f8c0eca3b4 [last accessed 23 February 2009].

National Institutes of Health (2000), *NIH Contributions to Pharmaceutical Development, Case Study Analysis of the Top-Selling Drugs,* Administrative Document, Office of Science Policy, February 2000.

Novartis (2009), *About NITD*, available at www.novartis.com/research/nitd/about.shtml [last accessed 28 February 2009].

Pécoul B., Chirac P., Trouiller P. & Pinel J. (1999), Access to Essential Drugs in Poor Countries – A Lost Battle?, *Journal of the American Medical Association* 281(4), pp. 361-67.

Public Citizen (2003), *America's Other Drug Problem: A Briefing Book on the Rx Drug Debate.* Washington, DC: Public Citizen, available at: http://www.citizen.org/documents/dbbapril.pdf [last accessed 23 February 2009].

Relman A. and Angell M. (2002), How the drug industry distorts medicine and politics – America's other drug problem, *The New Republic* December 16, pp. 27-41.

Ridley D., Grabowski H. and Moe J. (2006), "Developing drugs for developing countries", *Health Affairs* 25(2), pp. 313-324.

Sanders B. (2007), S.2210 The Medical Innovation Prize Act of 2007, available at: http://thomas.loc.gov/cgi-bin/query/z?c110:S.2210.IS: [last accessed 23 February 2009].

Sell S.K. (1998), *Power and Ideas: The North-South politics of intellectual property and antitrust.* New York: State University of New York Press.

Sobel D. (1995), *Longitude.* New York: Walker.

Sterckx S. (2007), "Lack of access to essential drugs: a story of continuing global failure, with particular attention to the role of patents", in Lenk C., Hoppe N. and Andorno R. (eds), *Ethics and Law of Intellectual Property – Current Problems in Politics, Science and Technology.* Aldershot: Ashgate, pp. 175-197.

Witty A. (2009), *Big Pharma as a catalyst for change,* Speech to Harvard Medical School on 13 February 2009. Full text on file with the authors. Summary available at: http://www.gsk.com/media/Witty-Harvard-Speech-Summary.pdf. [last accessed 23 February 2009]

World Trade Organization (1994), *Agreement on Trade-Related Aspects of Intellectual Property Rights, Annex 1C to the Marrakesh Agreement establishing the World Trade Organization*, Geneva: WTO, available at: http://www.wto.org/english/docs_e/legal_e/27-trips.pdf [last accessed 23 February 2009)

7

Human Rights and Transhumanism: Is a Humanist Transhumanism Possible?

Fernando Suárez Müller

Introduction

The object of this chapter is to clarify the relationship between two argumentative strategies in the field of eugenics: the humanist and the transhumanist. In order to clarify this relationship it will be necessary to differentiate not just between a humanist and a transhumanist perspective, but also between two main types of transhumanism. Transhumanists often refer to human rights by using humanitarian arguments in order to gain acceptance for their views. The question considered in this article concerns the use of these arguments in the setting of transhumanist ideals. The main arguments will be thoroughly discussed and evaluated. There seems to be a contradictory strategy because transhumanist ideals, which claim to transgress the domain of "conservative" humanist values, apparently need traditional humanist arguments. To clarify this apparent contradiction we will first have to make a distinction between two types of transhumanism. Secondly it will be necessary to demarcate two different types of eugenics. This will enable us to come to our third step which analyses the three main humanitarian arguments invoked in favour of transhumanism. The central humanist concepts concerned in these arguments are autonomy, authenticity and spirituality. To gain acceptance, transhumanists constantly use these three concepts which all refer to the notions of human dignity and human rights. By doing this they seem to be contributing to the idea that the Declaration of Human Rights needs an extension, because with the improvement of genetic interventions new rights seem to emerge. But by using this – typically humanist – language they fail to make clear why we should follow their ideals. Humanitarian arguments are at their best when justifying humanism and seem to be inappropriate for a transhumanist discourse.

Two main types of transhumanism: the conservative and the liberal one

In order to understand what the difference is between a conservative and a liberal approach to transhumanism, we need to refer to the work of Friedrich Nietzsche, who is often wrongly invoked as being the father of modern transhumanism.

"I teach to you the Overhuman. The human is something that shall be overcome. What have you done to overcome it?" These were the well known first words of Nietzsche's Zarathustra when he came from the mountain and, for the first time, met the people on the market-square.[1] What is the difference between Friedrich Nietzsche's *Übermensch* and Adolf Hitler's *Herrenmensch*? The most obvious answer to this question is that there is no eugenic program in Nietzsche's work. The German philosopher is talking about a rejuvenation of the soul. His message is mainly a spiritual and moralistic one. Zarathustra speaks as a religious prophet proposing a new system of values, which inverts the old, religious and humanitarian one. In the first chapter "On the Three Transformations" it becomes clear that what Nietzsche is aiming at is a transformation of morals. The camel, being a pack-beast, is not free. "There is much that is heavy for the spirit," says Nietzsche. The camel frees itself and becomes a lion which is strong because its spirit "seizes freedom for itself and becomes lord in its own desert". The lion is both the spirit of creation of new values and the spirit of destruction of the old ones. But to become a *Übermensch* a final transformation is still needed. The lion must become a child because the child is a new beginning, full of innocence and acceptance of life. Nietzsche's idea of the *Übermensch* is the result of a spiritual and moral search, although it is not spiritual and moral in the traditional sense.

Actual transhumanism seems to inherit some main ideas from Nietzsche. First of all it inherits the idea that the human is something

---

[1] Nietzsche, 1988, p. 14. I am using the translation of Parkes 2005, p. 11.

that must be overcome. The means to do that are biological (which was not quite what Nietzsche had in mind) but the end of such transformation concerns a moral and spiritual dimension which was the case in Nietzsche's philosophy. This moral and spiritual dimension can only be gained if old prejudices are destroyed. Nietzsche wanted to overcome humanism, but only – if we read him charitably – in order to make a higher form of spirituality and morality, a new form of humanism, possible. Transhumanism is making some similar claims today, which are to be realised by transforming our corporal constitution using biological and technical means. But Nietzsche was not interested in a biological transformation of humankind. His concern was a moral and spiritual idea. This is the main difference between his *Übermensch* and Hitler's *Herrenmensch*. Of course Hitler could not have conceived his racist eugenic program without presupposing a spiritual and moral transformation of inferior values. However, Hitler is not completely going astray in his interpretation of the *Übermensch* because passages can be found in Nietzsche that are clearly biological: "You have made your way from worm to human, and much in you is still a worm. Once you were apes, and even now the human being is still more of an ape than any ape is" (2005: 11). No doubt these biological passages are metaphorical, but the problem with metaphors is that they allow different interpretations and thus Hitler might have felt his ideas confirmed in Nietzsche's text. Not the term, but the concept of *Herrenmensch* was coined by Arthur de Gobineau. Here the main concept is a biological one transforming the idea proclaimed by Jacques Rousseau that all people have equal rights. The concept was taken up by Hitler who conceived the idea of systematically improving the German race. Hitler's "trans-humanism" – if we may call it that – was guided by an aesthetic idea of perfection that he saw represented in the Aryan race. Whereas Nietzsche wanted to turn back to the moral innocence of the child, Hitler conceived a system which had to turn back biologically to the pure and innocent beginnings of the German race: a race that at the beginning – like Nietzsche's child – was not corrupted by Christian or humanist ethics.

How different this is from the comic figure created by Jerry Siegel and Joe Shuster! The man of steel was a true humanist dedicating his strength to undo all attempts to neutralize the rights of man. Whereas Nietzsche's *Übermensch* and Hitler's *Herrenmensch* were anti-egalitarian ideas refuting the validity of humanitarian values, Superman is a defender of justice and human rights, first and foremost of which is the right of liberty. Superman however was born on Krypton and was not created by a eugenic program. But if he were, his image might have been used by the transhumanist movement for the front office of their sales department. In some respects Superman is a symbol for the message the transhumanist movement wants to transmit: he is a systematically biologically improved human being and this should be the main goal of real humanitarianism. Whereas the *conservative* transhumanism of Hitler wanted to change humankind in order to recreate an (illusionary) original and pure type of man, liberal transhumanism sells its ideas as serving egalitarian humanity, the artificial improvement of man being an implicit goal of human rights. To enhance our biological constitution is a matter of responding to the dignity of man which invites us to achieve a state of excellence.

Humanist and transhumanist eugenics

How does the contemporary sales department of the transhumanist movement function? First and foremost it functions by reducing the differences between the humanist and the transhumanist approach to eugenics. One of the strategies to reduce this difference is to leave the term 'transhumanism' largely undefined. What exactly is meant by 'transhumanism'? Nick Bostrom – one of the leading transhumanist thinkers – says transhumanism has one main idea: it wants to develop humanity's potential using science and technology. Its concern is to enhance humanity by technical improvements.[2]

This, of course, leaves largely undefined what transhumanism really is. In fact developing humanity's potential using science and technology has always been a cornerstone of humanism since the Renaissance. The real distinction between humanism and transhumanism is not to be found in this. It all depends on *how* science and technology are applied. Transhumanism wants to change the physical structure of mankind. As Nick Bostrom says:

---

[2] "I see my position as a conservative extension of traditional ethics and values to accommodate the possibility of human enhancement through technological means." See Bostrom, 2008, p. 113.

"Transhumanists view human nature as a work-in-progress."[3] And by 'nature' he means our biological nature: the way in which ev olution has made us. Transhumanism wants to "assume control over evolution".[4] This will mainly be done by genetic enhancement.[5]

Transhumanism is thus not just a matter of applying science and technology in order to develop humanity. It is an attempt to control the direction of evolution. Like Hitler, the liberal transhumanist wants to steer evolution, but they are going even further because they want to add something to the evolution of human kind that does not yet exist. *Conservative* transhumanism on the contrary wants to prevent the decay of humanity (or a part of it) by turning back to a mythological origin. In order to achieve that, a biological intervention is needed that steers evolution. In *liberal* transhumanism the main point is not to prevent any decay of the human race but to realise completely new forms of human life by systematically improving properties humanity already has and by creating whole faculties unknown to human kind till now.

Nick Bostrom argues that human dignity need not be affected by such changes. Human dignity can even be improved because human individuals will be able to (re)create themselves as they choose to be.[6] Transhuman eugenics is a human right because individual liberty includes the liberty to (re)create ourselves as we choose to be. This means, in fact, that the Declaration of Human Rights should be rewritten and extended some day in order to make it clear that steering evolution has become a matter of individual choice. However, the main practical question is: will each individual decide these enhancements for him- or herself or are we going to decide this for each other? Maybe in a transhumanist future there will be a realm of the human body on which decisions will be made by governments and another realm on which decisions will be fully private. Immortality, for example, although a private choice will probably need a governmental organisation, but changing one's nose will probably be organised privately. Governments will have to decide

---

[3] Bostrom, 2009, p. 345. See also www.nickbostrom.com/ethics/values.html.

[4] Bostrom, 2004, p. 369

[5] Bostrom, 2003, p. 493-506

[6] Bostrom, 2008, p. 13

which changes are allowed and which are not. There is a lot on practicability that *liberal* transhumanists should think about.

To say that an increase of humanity's potentials using science and technology, including eugenics, is possible, does not mean that all forms of eugenics are transhuman. A humanist eugenics is of course possible because we can think of improving our body genetically without changing the way evolution made us. If we can prevent diseases and deformations we do not add to evolution anything new. To be sick or to be healthy are both potentials of the human body as evolution has created it. If I prevent a disease I am just making a choice between two possibilities of being human. There is nothing transhuman in such a choice because I am not adding anything new to the human condition. The argument that we need transhumanism to end corporal suffering can therefore be denied. Psychic suffering can also be ended without needing transhumanism, although the question is to what extent this is desirable.

Humanitarian arguments used in favour of transhumanism

Do we buy the argument that transhumanism is an improvement on humanism? Can the rights of man best be realised by transforming our physical properties? To answer these questions we need to analyse the main arguments needed to come to such conclusions. Three arguments have been developed in this context: 1) the argument of autonomy (or liberty), 2) the argument of authenticity (or self-realisation) and 3) the argument of spirituality (or religion). In the next subsections, I will develop these arguments as strongly as I can in order to show that the counterarguments I am giving are not wholly unproblematic because they depend on metaphysical assumptions which may not be admitted by everyone. I think however that there is a *non sequitur* in all the argumentations enumerated above, because they do not succeed in showing that transhuman eugenics is necessary. The idea that a wise use of transhumanism would be beneficial to humanity can therefore be declined.

The argument of autonomy

Once we have clarified what transhumanism is and that humanist eugenics is something very different, we can move to an analysis of the main humanitarian argument used in favour of transhuman eugenics. Transhumanism is valued positively because it wants to overcome biologically-imposed limitations and imperfections of the human condition.[7] These limitations and imperfections restrict human dignity and human rights because they obstruct the full development of our individual liberty. If someone wants to be an astronomer he or she must have some talent for mathematics. Not having this talent means that there is something blocking his or her individual liberty.

In fact everything existing in space and time has its limitations. A question is whether all these limitations are imperfections. This is of course a normative issue that is not unimportant because it raises the question of who is going to decide about what. Even if we say that the final decision on such issues should be made by the individual, we can never be certain that the individual is not influenced by society. But if we take for granted that individuals are always somehow influenced by others this cannot be a major point in a counterargument. However, the wishes individuals have, if they are to be realised genetically, will have repercussions on society. Maybe it would be practical to have the eyes of an eagle or the strength of the Hulk? What about – as in the case of our untalented astronomer – improving our intelligence? Here a lot of questions arise: Are there different types of intelligence? Are certain types exclusive to other types? Should we improve them all or just some? Will I decide myself what kind of intelligence I will get? If we want to create exceptionally intelligent beings we are in fact also changing evolution, not on an individual level because being exceptionally bright is already a potential of mankind, but on a collective level because we are reducing the diversity and variability that society needs. It is exactly where the shoe pinches. If we allow everybody to change intelligence at will, the variability in society will decrease. But now suppose that in order to prevent such a degradation of society everybody were allowed to change their talents just one time. In order to prevent a

---

[7] See note 10. See also Bostrom, 2007, p. 107

decrease of diversity the wishes of some individuals will have to be denied so this would have to be organised by the state. Of course, this can be realised by a strong State, but what about the original point of unblocking our individual liberty? It is clear that free choice on these matters will not really lead to the development of full liberty.

It seems to be better to accept our limitations and talents as they are, as they have been potentiated by evolution. Humanist eugenics considers human beings to be "perfect" as they are: it only makes eugenic choices in cases belonging to the spectrum of evolutionarily-given human potentials. An individual with an innate disease blocking his/her talents for example would be a candidate for human eugenics. By "his/her talents" is meant an evolutionarily-possible potential: a mentally retarded person could be mathematically talented if he or she were not born with a disease. On the other hand it does not belong to human eugenics to, for example, make people see like an eagle. To see as eagles do, we already have developed the power of glasses. There is no need here to change the evolutionarily-established nature of man.

The counterargument developed above is in fact social and can be called the *argument of diversity*.[8] There is however another counterargument against the transhuman argument of autonomy which is related to autonomy itself. According to this counterargument, transhuman enhancement would be the defeat of autonomy. To understand this point, we first have to explain the distinction between the categories "transcendence" and "transgression". The traditional interpretation of human autonomy is related to the concept of transcendence. Humans are autonomous beings because they transcend the impositions of instinct. They develop their own goals in life, which shows that the mind commands the body.

Transhumanism conceives transcendence as transgression. Transcendence is the individual and autonomous effort to improve the pre-existing human potential. A transgression of the human condition is however something very different. Transcendence as it has been conceived by traditional religions and by humanism does not need a transgression of corporal properties. The gymnosophists

[8] See Habermas, 2001, pp. 105-114

of ancient Greece and the Indian yogis, for example, do not aim at a transgression of these properties. By showing that the mind is more powerful than the body they show that the human condition, as it is, already has the potential needed for tran-scendence.

In fact this is also the essence of sport since the ancient Greeks. It is not just to show who is best, but to show that the will, the mind, as part of the human condition, is in charge. It is a demonstration of the triumph of the spirit over the body. That is the reason why doping is so unfair. It is not just unfair in relation to others but also in relation to oneself. Doping is the defeat of autonomy. And very similar to doping is transhuman eugenics. It is here not the individual himself whose will triumphs over his body. External elements like drugs or surgical and genetic interventions are now doing the job. These interventions not only endanger the autonomy of the individual because the decision to enhance my cells may be made by others; they do not just *endanger* autonomy, in fact they ruin it and make any achievement following the artificial enhancement less valuable. If I improve my brain cells in order to be brighter and more creative this is not an achievement of the self, it is not an autonomous accomplishment. The argument of autonomy used in the context of transhumanism is digging its own grave.

It is true that human knowledge increases using instruments like telescopes, computers and books. But as long as these are used as external instruments which can be taken up and left aside at will, there is no transgression of the human condition. If we, however, improve human vision genetically in order to make telescopes superfluous, this would clearly be a transgression. As long as we can use external instruments there is no need to incorporate them in our body. Now, what about having chips implanted in our brains in order to have direct access to encyclopaedic information or in order to improve memory? It is possible to substitute a damaged neuron by a chip and maybe this will contribute to the prevention of Alzheimer's. But this is not different from any other artificial prosthesis replacing a missing or malfunctioning body part. The autonomy of the individual is not endangered by these devices, because they only have a substituting function. Other devices like encyclopaedic memory sticks that could be implanted in the brain are trickier. This is clearly a case of transhuman technology. In this case autonomy depends on the ability the individual has to distance herself from the stick. If the stick has become an organic part of the brain the achievements of an individual are artificial and we may have serious doubts about her autonomy. The main question concerning these types of transhuman transformations of the human body is in fact: Do we really need these transformations? Why should we integrate into our body instruments enhancing the singular possibilities given to everyone by evolution when it is also possible to keep these devices outside the body?

In fact the real difference between the humanist and the transhumanist eugenics and technology is based on a different metaphysical worldview. To the transhumanists, the self is just an epiphenomenon of matter and therefore a spiritual striving of the self – autonomy – can only be conceived as the result of a material constitution. If it is just that, there can be no impediment to the recreation of matter by technological and genetic means. The epiphenomenon resulting from the newly constituted brain or body will be considered to be as autonomous as the epiphenomenon of the old body. To the humanists on the contrary, the self cannot just be an epiphenomenon of matter. The self has a dignity and existence of its own. We cannot discuss this point in further detail because it would lead us to the endless discussions of contemporary philosophy of mind. We just want to make clear that the humanist *counterargument of autonomy* implies a metaphysics which will probably not be accepted by transhumanists.

The argument of authenticity

Another humanist argument used by transhumanists is the argument of authenticity or self-realisation. Transhuman eugenics and technology are valued positively because they make individual self-realisation possible. There is a strong connexion with the argument of autonomy: In fact the argument of authenticity presupposes the argument of autonomy. If we take again the case of the untalented astronomer mentioned above, the lack of a mathematical talent is not just blocking his or her individual liberty, it is also impeding his or her self-realisation. Transhuman technological interventions can make this self-realisation possible and thus be a means for people to live a life in full authenticity.

This is in fact one of the points referred to by Aubrey de Grey in order to sell his ideas on corporal immortality.[9] Life is too short to realise all the things we want to develop. We always have to make difficult choices because we run out of time. If we could extend life at will we could develop our personality in many directions, we could become experts in mathematics and be at the same time incredible acrobats, experienced surgeons and virtuoso musicians. There is no question here of artificially adding talents or properties in order to become more autonomous. The point is to be enabled to develop the talents we already have and that cannot adequately grow to real fulfilment because our lifespan is too short.

In a strict sense immortality is of course impossible for material beings. They are bound to physical forces: if the universe collapses no transhumanist is going to stop this. The idea of immortality is in fact relative and not truly attainable. What really is at stake here is longevity and not real immortality. I already argued elsewhere that such unlimited longevity is morally undesirable because it either leads to overpopulation and all the ecological and social problems related to that, or it leads to a society that gives up future generations.[10] These consequences would make a choice for unlimited longevity an egocentric and immoral one. To be able to sacrifice oneself for others has always been a distinctive property of moral action. In a future in which unlimited longevity is possible, to die naturally would therefore become a moral decision.

It seems at first that the transhuman argument of authenticity can be separated from the idea of corporal immortality although it will always refer to the question of longevity and rejuvenation. We could, for instance, decide only to extend life and to rejuvenate our body till we have accomplished all our desires. We could decide to die once we have fully developed and exerted all our personal talents. A choice for longevity would then not be an immoral one, because we do not give up future generations. The question that rises would be: is such a moment of total fulfilment of personal wishes really possible in earthly life? Desire is probably limitless and could reproduce itself in infinity, which means that longevity would be identical with corporal

---

[9] We need immortality and childlessness because "there are plenty of things to do with our time." See De Grey, 2004, p. 238; see also: De Grey, 2005, p. 659-663

[10] De Grey, 2007, p. 225-239

immortality. At the end we will have to acknowledge that the transhuman argument of authenticity cannot be separated from the idea of corporal immortality and that it will remain connected to the suspicion of being immoral.

This counterargument can be called the *argument of generations*. Like the *argument* of diversity it is mainly a social argument. Next to this argument there is however a counterargument against the argument of authenticity based on the idea of authenticity itself. We already have made a distinction between a transhumanist and a humanist argument of autonomy. The same can be done for the argument of authenticity. The humanist version of this argument contradicts and neutralizes the transhumanist one.

The main point of the humanist argument of authenticity is that self-realisation is not about the realisation of all our wishes. Self-realisation is a question of searching the inner core of what really moves us, and this search must be done within the framework of morality. This inner core is what has been called the essence of man and this essence has a general part – referring to us as human beings – and an individual part – referring to us as persons. Self-realisation therefore means a concentration on the most pressing talents. The idea of a selection belongs to the search of the self: we select and concentrate on those things which really move our soul. Realising oneself is to search a hierarchy of values. If there were no selective pressure we would not discover our inner core and our soul would be scattered amongst the content of the world. To discover this inner core our normal human lifespan is enough. The history of mankind shows that people can find their own ways. It shows that talents can be developed during the course of an average lifespan. It shows that people can even reach a peak and then afterwards decide to develop another talent or develop variations of already known talents. The argument of authenticity used by transhumanists is therefore not really compelling. There is no need to change our biological nature in order to make authenticity possible.

I think it is true that humanity has a natural desire for immortality. In fact it is not just humanity but life itself that has this desire. Through evolution life has made the choice of satisfying this desire by reproduction and not by individual "immortality". But man has attained such a dignity that we become conscious of this old desire. We become conscious of the imperfection of ourselves and the world. We know an inner ideal space in which things are different.

That this space refers to a dimension beyond physical reality is the core idea of religion. That this inner dimension may inspire us to improve our personal, social and physical world is understandable and it is what motivates "progress". But mankind, conscious of our dignity, knows that all the changes we make must be improvements in a moral sense. Morality not only constitutes a necessary limitation, it should constitute the glue sticking together all our actions and thoughts. Once we have recognised (through the argument of generations) that corporal "immortality" cannot be in accordance with morality, we should abandon this idea and concentrate on the possibility of an ideal space beyond our physical reality.

The history of mankind shows that religious and spiritual people do not need unlimited longevity. They prefer to die in order to leave the world to others and to obtain, by one way or another, a perfect state of being after death. But of course all this presupposes a worldview which is miles away from the materialistic perspective of transhumanism.

The argument of spirituality

In the course of the argument of authenticity, as we just came to see, we find the argument of spirituality. This is an argument that has not been developed by the transhumanist movement itself, because its worldview is mainly materialistic, but comes from some types of holistic thinking based on the pantheistic views of Ken Wilber. Here, transhuman eugenics and technology are valued positively because they may make a spiritual or mystical development of individuals possible. On the one hand science could make our intelligence grow up to the level of savants, saints and mystics. On the other hand it could make corporal immortality possible, so that our intellectual capacities may not be restrained by mortality. Transhumanist eugenics – although based on a materialistic worldview – can transform humanity into a spiritual society. Such an argument has been considered by Tatjana Kochetkova.[11]

I fully agree with the Wilber-inspired critique of transhumanism as being a form of materialist reductionism (a reduction of the mind to material processes), but I do not think that an alliance between

---

[11] See the next chapter

integral or holistic thinking and transhumanism is desirable. To escape reductionism Wilber has conceived a rather schematic way of approaching things, starting from a schema of being (a quadrant) that is not very easy to grasp.[12] A lot of ontological correlations offered by the quadrant are very confusing. In this schema atoms are correlated with galaxies and molecules with planets, which of course is confusing because both galaxies and planets are made of molecules *and* atoms. The intention of this quadrant is to create an essential connection between the material and the spiritual world. For instance, when speaking about the brain a holistic approach to the brain cannot just be neurophysiological, it also has to study (as belonging to the same level) our ability to use concepts, the social associations made possible by these concepts and the myths conceived with them. Holism in this sense is a kind of multidisciplinary approach guided by a scale of hierarchical levels of consciousness. For instance, atoms have their own level of consciousness; molecules have a larger one. And organisms and mankind, of course, have higher levels of consciousness. Wilber does not believe that our mind is an effect of the brain. The spiritual side of being has its own ontological existence. However, the development of the brain in the material world gains access to a higher stage of spiritual evolution.

One remarkable thing about Wilber's quadrant is that the stages of consciousness all depend on material evolutions: the neural cord for instance makes perception possible and the cortex conceptual thought. If we follow this scheme we can only achieve higher forms of consciousness when there has been a change on the material level. The scheme compels us to think that more and more material mutations are needed in order to make higher forms of consciousness possible. Here holism and transhumanism shake hands. Such an alliance between holism and transhumanism is possible in every form of pantheism, because here spirituality is always being correlated to materiality.

But do we need an unlimited longevity or superhuman brains in order to become religious or spiritual? Visionary intelligence or moral growth does not depend on how long we live, and we do not need to be "immortal" to become wise, moral or spiritual beings. We can

---

[12] Wilber 2007, p. 110-111.

obtain all this within the limits of our average lives. We are not going to realise ourselves more or to become more spiritual just by adding years to life. That spiritual fulfilment is possible during a normal lifespan has been demonstrated by all religious mystics of the past who never achieved nor needed corporal immortality. The same applies to the enlargement of our brains. We do not need more brain power to become religious or spiritual. Of course it is known that drugs can alter our vision of the world. For the purposes of our argument, changing the capacities of our brains by eugenics is essentially the same as using drugs. In some sense we can say that the use of drugs diminishes the value of spirituality because we are not certain anymore whether the visionary state achieved is real. The problem of "religious experiences" is always that they may be the illusionary result of chemical circumstances. That is why religion cannot rely on "spiritual experiences" alone. Experiences in general are not always trustworthy. Religiousness must therefore also be accompanied by rational reasoning.

Final remarks

Do we really need to change human rights by inserting paragraphs in the Declaration of Human Rights claiming our right to rejuvenation, to corporal "immortality", to have a free choice to fly as Superman or to be as strong as the Hulk? Transhumanism is not just a transgression of evolution, it is also a transgression of morality on three levels:

1) Mankind is an end to itself, which means that it must develop its own potential autonomously. Transhuman manipulations are equivalent to the use of doping: superhuman growth is not the result of one's own labour.

2) Mankind is an end for itself, which means that life is a workplace where a lifestyle must be found unifying universal moral principles and individual desires. Most religions tell us that this implies an internalisation of love into one's own being, which is nothing more than searching out this love inside us, because it is always already there. If this is the case there is no need for transhuman manipulations. Such manipulations would be a sign that we are searching in the wrong place, that we are substituting the art of morals by the art of appearances.

3) Mankind is an end to or the point of human rights. These, although universal, are based on our human condition. If we change the shape of humans we undermine the basis of these very rights. Transhuman manipulations claiming a human motivation transform what makes this motivation necessary. Transhumanism eats away the foundation of human rights and of its own inspiration.

Eugenics must not be reduced to the narrow focus of scientism and materialism. Modern liberal transhumanism is, in fact, an example of such a contracted perspective. It has to use traditional humanist arguments – which then in a natural way fail to make their point – because it has no arguments of its own. The approach to eugenics must be holistic in the genuine sense of the word: it must be wide and not reduced to the scientific perspective. Cultural, social, aesthetic, ethical and philosophical perspectives should be included. Only in this sense is a *lingua democratica* of eugenics possible. But just as democracy cannot function without agreeing on a constitution, on some elemental values, a *lingua democratica* is not going to function without a *lingua ethica*. Just as democracy presupposes a humanist constitution, there must also be fundamental humanistic values making the *lingua democratica* possible. It is the task of ethics to specify the hierarchy of values that will determine which action, which research program and which technological invention must be implemented first. It is all a question of priorities, and transhumanist eugenics, I would say, does not deserve this priority.

Literature

Bostrom, N. (2004). "The Future of Human Evolution" in: Tandy, 2004: 339-371.

——— (2007). "Why I Want to Be a Posthuman When I Grow Up", in: Gordijn/Chadwick, 107-137.

——— (2003). "Human Genetic Enhancements: A Transhumanist Perspective", *Journal of Value Inquiry*, 37/4, 493-506.

——— (2008). "Dignity and Enhancement", in: *Human Dignity and Bioethics: Essays Commissioned by the President's Council on Bioethics*, 173-206.

——— (2009). "Transhumanist FAQ", in: Kaplan, 345-361.

De Grey, A. (2004). "Aging, childlessness or overpopulation: the future's right to choose", in: *Rejuvenation Research*, 7/4, 238.

——— (2005). "Life extension, human rights, and the rational refinement of repugnance", *Journal of Medical Ethics*, 31/11, 659-663.

——— (2007). "On Futuristic Gerontology", *International Journal of Applied Philosophy*, 21/2, 225-239.

Gordijn, B. & Chadwick, R. (2007). *Medical Enhancement and Posthumanity*. New York: Springer

Habermas, J. (2001). *Die Zukunft der menschlichen Natur. Auf dem Weg zu einer liberalen Eugenik?*, Frankfurt: Suhrkamp

Kaplan, D. (Ed.) (2009). *Readings in the Philosophy of Technology*, Plymouth/New York: Rowman & Littlefield

Tandy, C. (Ed.) (2004). *Death and Anti-Death: Two Hundred Years After Kant, Fifty Years After Turing*, Palo Alto: Ria University Press

Nietzsche, F. (1988). *Also sprach Zarathustra*, vol. 4, Berlin: de Gruyter

——— (2005). *Thus Spoke Zarathustra*, transl. by Graham Parkes, Oxford: Oxford University press

Wilber, K (2007). *A Brief History of Everything*, Boston: Shambhala

8

The Spiritual Dimension of Human Enhancement: Remarks concerning the Discussion about Immortality

Tatjana Kochetkova

> ...there is such a thing as perfection...and our purpose for living is to find that perfection and show it forth....Each of us is in truth an unlimited idea of freedom. Everything that limits us we have to put aside.
>
> *Richard Bach*

Introduction

Nobody involved with the evolution of consciousness could possibly remain indifferent to such a recent intellectual trend as transhumanism. "The world's most dangerous idea," in the eyes of its critics, transhumanism is also claimed, in Bailey's words, to epitomize the most courageous, daring, idealistic and imaginative human aspirations.[1] One can define transhumanism as the trend which strives to "assume control over biological evolution" by means of genomics. One still has to add that some transhumanists rely on non-biological technologies as cybernetics, and do not focus on human DNA.[2]

The two cornerstones of transhumanism are, firstly, that humanity should develop its potential of unlimited perfection, and secondly, that science and technology are the way to attain this goal. This first statement obviously coincides with millennia of the old spiritual

---

[1] Bailey, Ronald, Transhumanism: The Most Dangerous Idea? Why striving to be more than human is human, at http://www.reason.com/rb/rb082504.shtml

[2] Bostrom, N., "Human Genetic Enhancements: A Transhumanist Perspective": www.nickbostrom.com/ethics/genetic.html

search, which calls for an analysis of the compatibility between transhumanism and the spiritual quest with respect to the goal of overcoming the limits of the human condition. Let us look at the difference of perspectives from which these two approaches pursue their common goal.

Transhumanism, having arisen from the cyberculture of the late 20th century, is new, enthusiastic and appealing. Critics complain about its "nerdy" aura, due to a large intellectual component, but *per se*, this can hardly be considered a disadvantage: many discoverers and inventors were ridiculed as "nerds." At its origins one finds Julian Huxley, a brother of Aldous Huxley, and later Fereidoun M. Esfandiary, who used the pseudonym "FM-2030." Among transhumanism's constitutive essays is Max More's "On Becoming Posthuman", which emphasizes the goals of achieving unlimited existence and enhancing all essential human capacities.[3] The transhumanist movement sprang from the work of philosophers Nick Bostrom and David Pearce, who founded the World Transhumanist Association.

Transhumanists consider the current form of humanity a transitional stage towards other, enhanced states. To accomplish this transition, transhumanists rely on Enlightenment rationality as the major means to improve the human condition. The association with the Enlightenment leads to most transhumanists being agnostic or atheistic, although other worldviews are also represented, such as Buddhism, or non-orthodox Christianity. Even though the transhumanists' critique of religion (see e.g. the essay by Max More mentioned above) misses a crucial point by mostly targeting those aspects of religion that have little to do with true spirituality, their enthusiasm and adventurousness certainly seem to breathe new life into the more attractive sides of the Enlightenment. Transhumanism relies on research in areas as diverse as human genetic modification, cryonics, nanotechnology, cybernetics, space colonization and others, arguing that they will eventually lead to the means whereby the human condition can be overcome.[4]

---

[3] More, Max, "On Becoming Posthuman", 1994, at http://www.maxmore.com/becoming.htm

[4] The techniques we have mentioned have a potential to variously enhance human capacities, including the achievement of physical immortality. Thus, gene

The transhumanist goal of transcending the merely human is frequently misunderstood. It can seem shocking to somebody educated in the classical Western philosophical tradition, in which human beings are held in the highest regard. In Kantian terms, we think that humans are ends in themselves, the highest perfection on earth, and that humanity is something to be maintained at all costs. A challenge to this can sound like technocentric arrogance, and this impression provides some insight into the motivations of Fukuyama's critique of transhumanism.[5] However, on closer examination, one can see that transhumanists don't actually challenge human values or consciousness as such, but only human limits, imperfections and suffering. Are these not obstacles that we have always struggled to overcome? Does this wish not come from the core of the human spirit? A convinced spiritualist like the romantic philosopher Vladimir Solov'ëv could also be considered a transhumanist: he claimed that the development of consciousness, transcending limitations and defeating death must have priority for human action.

Secular objections to transhumanism

Transhumanism is an object of various criticisms and discussion.[6] Some of the objections to transhumanism rest on a misleading perception of it as inhuman, while others highlight serious practical

---

transfer allows correcting the work of certain genes of adults, and by genetic modification of the human germline it is possible to modify (inheritable) genes of embryos. It is probable that genetic modification can postpone, reverse, or arrest the process of aging. Cryonics is a technology that currently uses low temperatures to preserve human organs for transplantation. Some transhumanists practice the immediate freezing of the human body right after death at a very low temperature in the hope to preserve it until the time when it may become possible to revitalize the body. Nanotechnology offers a possibility of correcting the work of specific genes or organs by using nanobots, and this also offers various possibilities for enhancing human capacities. Transhumanist thinking includes speculations on space colonization and cosmic expansion of trans/humanity in the future. Some transhumanists consider the possibility of transferring human mind into computer hardware thus overcoming mortality.

[5] Fukuyama, Francis, "Transhumanism", *Foreign Policy*, September/October 2004

[6] I express my gratitude to Fernando Suárez Müller for his careful criticism and discussion in the foregoing chapter.

problems. Max Dublin doubts the accuracy of transhumanist predictions concerning future radical changes, on the grounds that similar predictions in the past (Marxism, Millenarianism) have turned out to be wrong.[7]

Other critics find such a transformation possible, but undesirable for ethical reasons, or too risky to be worthwhile. They use, for instance, the "Terminator" argument (the technological advances suggested by transhumanists may cause human extinction), the "Brave New World" argument (new technologies may lead to the end of political and personal freedom), or the "Frankenstein" argument (new technologies may have harmful consequences and should be treated with caution). To continue, the "Enough" argument claims that changes in the fundamental conditions of human life are undesirable and may diminish our humanness; the "Gattaca" argument warns that new technologies may widen the gap between the rich and the poor.[8] These dangers are certainly all real, but still avoidable. In my opinion, it is not the risk of transhumanism, but how well this trend reflects and answers human aspirations, that will decide its future. It seems that overcoming oneself is a deep human desire without which we can hardly exist; and this seems to be also important for transhumanism.

Some critiques of transhumanism claim that the attempt to control human biological evolution, particularly positive eugenics, is unethical as such. However, whether such an attempt is ethical or not, shall be judged according to its goals and means. If the goal is, as in case of some transhumanists, to improve the condition of humankind, to alleviate suffering and increase the possibility for realizing human potential, and if it is done in a way that respects human rights and individual freedom of choice, then one cannot dismiss transhumanism without further discussion.

Another argument claims that enhancing intelligence would reduce the diversity and variability, needed by society. However, the same argument was used in the 19th century against universal secondary education: it was claimed to reduce the diversity in intelligence, needed by the society. But, as we see, the society was

---

[7] Dublin, Max, *Futurehype: The Tyranny of Prophecy*, Penguin Books, 1992

[8] McKibben, Bill, *Enough: Staying Human in an Engineered Age*, Times Books, 2003; see also McKibben, Bill, *The End of Nature,* Peter Smith Pub. 2001

flexible enough to accommodate this change. In fact, enhancing human intelligence is what the education system was doing through the entire history, and genetic improvement would do it by different means but with the same goal.

Another practical argument claims that unlimited longevity will lead to either overpopulation or the renouncement of future generations. The obvious solution that springs to mind is the colonization of the outer space. The technology that can achieve unlimited longevity would most probably also be capable of providing ways of space colonization. However, according to contemporary astrophysics, the physical universe is limited. Therefore, any colonization of the outer space will still lead to overpopulation in the very long run. However, taken into account the enormous size of the universe, such overpopulation may take millions of years, and before that new solutions might be found.

Concerning the ethical side, it is argued that improving the physical performance and reducing the physical vulnerability of human beings will still make it impossible to demonstrate the triumph of spirit over body in sport, similar to doping.[9] However, any enhancement of physical performance can be only relative and limited, while the relation between body and mind will stay the same: in each physical performance, be it sport or arts, the mind will always lead no matter how advanced the body might become. Physical enhancement of athletes would place sport competition at another scale of speed or strength, but it cannot change the nature of the mind-body unity that plays a crucial role in each competition.

The goal of a spiritual quest

Let us look at the goal of a spiritual quest, as distinct from an established religion. Under the spiritual quest here I mean the philosophical tradition which aims at overcoming the limitations of the human condition. The human condition is characterized by the disparity between the external, material appearance and the human,

[9] The unfairness of the use of doping rest on the fact that not every sportsman uses it; but if everybody can use safe doping, the competition will remain fair, just occur in other circumstances.

internal dimension. Internally there is an intuition of infinity that is an important part of our inner experience of life, at any age, in any place. The disparity between the human condition seen from outside and seen from inside is the source for spiritual inquiry from most ancient times. This inquiry is reflected in many world masterpieces, from the *Epic of Gilgamesh* and the wisdom of the *Upanishads* to Goethe's *Faust* and Nietzsche's *Zarathustra.* King Gilgamesh's legendary quest for eternal life failed, but all the intervening generations started it over, again and again. Among the representatives of this quest is the great part of the philosophical tradition, also sometimes called 'perennial philosophy' – from Plato and Neo-Platonists to Nicholas of Cusa, Pico della Mirandola, Novalis, Goethe, Schelling, Emerson, Solov'ëv and currently Wilber and transpersonal thinking.

There are three general ways to deal with the dilemma between inner infinity and external limitation. One is to accept the human condition in the way it appears to be from the outside, in which case the inner intuition is forgotten. The second is to flee from this world into the realm of Spirit. The third is to take this feeling seriously and investigate how this disparity appeared. This latter view solves our dilemma by taking a dynamic approach to the human condition. It suggests that humans are indeed something special. Behind the fragile and passing human appearance there is a deeper reality: Atman, Self, Psyche.

This quest can be interpreted as a search for authentic spirituality, for authentic transcendence. One of its possible embodiments is the quest for *self-realization.* Spirituality always implies transcendence, going beyond what is merely given. This is also the goal of self-realization, as can be seen by comparing this idea in humanistic psychology and the related ideas of individuation in psychoanalysis, the present-day growing interest in self-knowledge, ancient mystical traditions, and yoga. This interest does not always take a sophisticated form: examples are the popular "Superman" and other masked superheroes in popular culture, or that of transhumanist cyberculture's proposal of transplanting human consciousness into computers. Despite their many problems, the ultimate goal of these approaches is still the *overcoming of human limits*, which is what we term *self-realization.*

The content of self-realization can be better clarified by means of comparing it to another different but related concept of *self-actualization.* Introduced by Maslow, self-actualization denotes a stage in individual development at which the lower or basic needs are gratified and the "higher" (*meta-*)needs, like truth, creativity or justice, drive individual activity. Self-actualizers, contrary to what a superficial understanding might suggest, are the opposite of selfish people, because they are motivated by goals beyond themselves. They may not yet be self-realized, but they are clearly on the path towards this goal.

The humanistic idea of self-actualization became received knowledge in psychological science, although in a limited version. Today's scholars, both psychologists and philosophers, frequently think that this is all that it is about, and do not take it further towards self-realization. To them, self-realization is at best a matter of individual concern, arbitrary and subjective, concerning personal happiness rather than a profound quest within human nature for its hidden potential. Most researchers take it for granted as a mere subjective desire without examining its possibilities and what it tells us about human beings. However, Maslow himself thought that there are still farther reaches of human nature beyond self-actualization. In contrast to scholars who see self-realization as a purely subjective desire of some individuals, many of today's broader cultural trends consider it as expressing common human potential.

The quest for self-realization shall be understood as the deepest demand of human nature and the driving impulse in human evolution, an on-going process. This deeper perspective on self-realization relates to the historical potential of the human race. It assumes that human consciousness evolves through history, and that even the pinnacle of its present achievements is only a passing stage. What is to be realized is the infinite human potential that transcends space and time and differs from the limited individual ego. The realization of this self is a more radical transcendence of the limitations of human existence than Maslow's self-actualization. On the scale of human development, absolute self-realization relates to contemporary "normal" consciousness more or less like a mature genius does to a new-born baby. Believing in infinite human potential, the Russian philosopher Vladimir Solov'ëv argued that, just as the incapacity of a new-born to speak does not place language beyond human capacity, the failure of all previous philosophy to find an authentic reality does not in itself deny human potential the ability to acquire this knowledge. Another kindred spirit whose views were quite close to Solov'ëv's opinion on absolute self-realization was Teilhard de Chardin, with his concept of Omega consciousness.

Transhumanism and the exterior dimension of human existence

Let us return to transhumanism. It shares with the spiritual quest the goal of continuing the evolution of human nature, while transhumanism relies generally on the Enlightenment worldview and science and technology as the means to achieve this goal. Still, as I will attempt to show, the conventional techno-transhumanism is partial and one-sided from the integral point of view, because it focuses exclusively on the exterior dimensions of existence. Here we apply Ken Wilber's diagram (exterior-individual and exterior-collective) to explicate this. Thus, according to Wilber's scheme, each social entity can be regarded in four dimensions of existence. His approach was reflected in a scheme that he calls "the four quadrants" of existence. The scheme includes over two hundred concepts used in various branches of knowledge, from the natural sciences to the humanities and mystical philosophy.

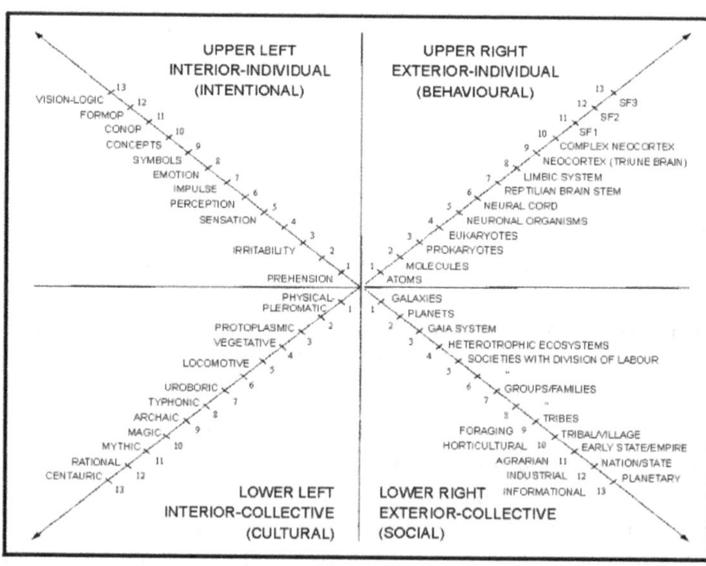

Figure 1 Ken Wilber's four quadrants scheme

Wilber's four quadrants refer to inner experience, both as individual intentionality and as cultural symbols, and to external experience, physical as well as social. Wilber's integral approach requires correlating the phenomena of the same developmental level at each of the quadrants to see the whole of an event. For instance, having a complex neo-cortex corresponds to the ability to operate with concepts, and it corresponds to the agrarian tribal organization at the social level and to magic thinking (archaic mythologies) at the cultural level. These four dimensions have to be regarded as expressions of the same event, in the integral framework. In other words, we must "track simultaneously" all four dimensions we study for each phenomenon.[10]

Specifically, the suggested methodology includes ways appropriate to all valid knowledge (injunction, apprehension, confirmation) and a different type of validity claim for each quadrant (behavioural quadrant — prepositional truth, intentional quadrant — subjective truthfulness, cultural quadrant — cultural meaning, social quadrant — functional fit). As one may notice, he incorporates here Habermas' theory of communication. The *integral* aspect consists of keeping track of all correlations between levels and quadrants, which Wilber considers the way to an authentic theory of consciousness. However, none of the quadrants can be reduced to another one without distortion.

In categories of the integral scheme, transhumanism believes that the interior dimension is derived from the exterior dimension, thus presenting a case of what Wilber calls "flatland holism" or "subtle reductionism". Subtle reductionism happens without paying attention to the independent and irreducible character of their interior counterparts. By focusing on humans only as bodily, biological beings, mainstream transhumanism tends to forget about the existential side of human life. Along with a perfect or enhanced body, a human being also needs the capacity to have a meaningful existence, and this is not a question of technology. This omission does not make transhumanism hopeless, but it certainly brings up important gaps. The result is a one-sided picture of reality, grossly

---

[10] See for Wilber's four squadrant scheme: www.wilber.shambhala.com

underplaying the inner dimensions. Still, this does not deny the useful and valuable character of the transhumanist undertaking.

The two sides of immortality: physical carrier and meaningful existence

The transhumanists' partial but worthwhile approach appears most vividly in the discussion on one of their objectives: immortality. While claiming that overcoming death is necessary for human fulfilment, some transhumanists direct their efforts at creating a material carrier for human consciousness with, in principle, an unending existence. The physical nature of this carrier, however, is secondary: it could be an ordinary body reinforced with nanobots, a genetically enhanced body, iron/silicon hardware, or some other, yet unknown substrate. This approach assumes that human identity and consciousness are recorded in brain chemicals, taking for granted that consciousness is naturally mortal and denying the existence of an eternal higher self or Atman. One can argue that this is a "materialist" version of holism. This might suggest that transhumanism is incompatible with the spiritual quest, but, as we shall see, it does not seem so in the final account.

Firstly, embodied immortality and spiritual quests are compatible, as one can see in the case of yogis or of Christian mystics. Many centuries B.C. yogis had already been pursuing the same endeavour, on the evidence of Mircea Eliade and other historians of yoga.[11] At first it may sound paradoxical that yogis, who with their beliefs about spiritual immortality should worry least of all about bodily immortality, were the first to search for it. Their argument was that embodied existence in the human form presents far better conditions for the development of consciousness and spiritual enlightenment than disembodied or animal existence. They wanted to fully use the fortunate chance of being born human.

Also, for transcendentalist philosophers, like Solov'ëv or Kant, the obstacle for the attainment of the superhuman level of consciousness lies not in the form of the human body but only in the fact of being

---

[11] Eliade, Mircea, *Yoga. Immortality and Freedom* (in Russian), St. Petersburg: Lan', 1999

mortal: the essential requirement for realizing human nature is immortality. In contrast to Kant, Solov'ëv found that immortality of the soul is not sufficient for achieving the task of human existence because it cannot remove the tragedy of death, and therefore, the immortality of one's actual body is required. This means the immortality of the complete individual as a unity of spirit, soul and body. At the same time, the immortality of the individual cannot be reconciled with the obvious emptiness and vanity of our ordinary existence. In its own turn, immortality requires infinite perfection and never-ending personal growth, which is the realization of the eternal essence, revealed by love, in the empirical personality here and now.[12] Transhumanism addresses the side of creating an immortal body but leaves unattended the aspect of the meaning of such infinite existence.

The human project: the existential side of immortality

The basic divergence between transhumanists and supporters of the spiritual quest is that the former mostly think that immortality only consists of the capacity of making one's body to exist unlimitedly. A fundamental problem, however, is that such unlimited existence is *a priori* not clearly compatible with human consciousness and values; the consequences for humanity may be negative. "I'm only 48 and already periodically bored", claims Leslie Walker.[13] Her argument finds support with McCurdy, who believes that "part of what makes life a great adventure is knowledge it will end" and concludes that unlimited life extension will make us inhuman.

These arguments rightly point out that the obstacles to embodied immortality lie not only in the fragility of the human body, but also in

---

[12] As one can conclude from this, in Solov'ëv's system, death is not a natural destination of the human being but only a consequence of our imperfect empirical condition. I would further suppose that overcoming imperfection also means overcoming death. The idea of achieving immortality for all humankind was common to Solov'ëv and Fyodorov, who suggested a special project for this. In: Solov'ëv, V.S., *Smysl žizni* [*The Meaning of Love*], in: *Sobranie Sočinenij (Collected Works)* Brussels, V. 7, p. 31

[13] Walker, Leslie, 'Kurzweil's Quest For Eternal Youth Sets Group Abuzz', at http://www.washingtonpost.com/wp-dyn/articles/A11564-2004Oct6_2.html

consciousness itself. Ordinary human consciousness is still existentially closed and therefore finite. The way to change this lies in psycho-spiritual growth, which cannot be accomplished by any technical means. The centre of gravity still lies in the psycho-spiritual evolution, in the inner dimension, in intentionality. For infinite embodied existence, *das Sein* shall open itself to the unlimited sources of meaning in the here-and-now, a transformation that transcendentalists, like Schelling, Emerson, Solov'ëv or Wilber are concerned with (Solov'ëv suggested erotic love as a doorway to it). This is the goal of the spiritual quest. Humans, in contrast to animals, are their own project. One can talk about the 'human project':

> It is natural for a man to desire to become better and more than he is now, it is *natural* to tend toward the superhuman ideal. If he *really* wants this, then he can, and if he can, he also must.[14] ....Man, even though he is also a creation of the reality, that is given to him, that existed before him, at the same time, he can affect this reality from inside, and consequently, this reality to one or the other extent *is that which he himself does* – does more or less obviously as a *collective* being, and less obvious, but just as surely, as an *individual person*.[15]

The target of this project is the internal growth of human nature as a part of *cosmic growth*, manifested in the increasing complex perfection of organic forms. Even though Solov'ëv found that the human body should not be changed, he still claimed that human perfection requires embodied immortality, as the life of a disembodied soul was insufficient for perfection through love. But then, how can one think of embodied immortal existence without transforming the human body in some way? Thus, the transgression of human biology towards physical immortality was implied in his project as well.

Among the objectives of this growth is overcoming the various inner reasons for human suffering. These are the internal splits: Freudian complexes, *Angst*, not being one with oneself and one's body – splits within the mind, splits between body and mind, and also splits between humans and the universe, and so on. The separate self

---

[14] Vladimir Solov'ëv, *Lecturers on Divine Humanity*; tr. B. Jakim, Lindisfarne Press: Hudson, New York, 1995

[15] Ibidem

creates its own hell. This hell can be overcome by inner development, in the transition from Mental Ego to Centaur, and from Centaur to higher levels of evolution. There are examples of people who did this, and humanistic psychology brings out arguments in favor of its feasibility.

The umbrella of the humanistic approach covers various practices, such as Gestalt psychology, Reichian bodywork, encounter groups, psychodrama, psychosynthesis, or neurofeedback. They all appear as divergent and, with their specific practices, all target different aspects of the human being, aiming at its development and integration into the whole of the personality. Yet, all these different methods seem to pursue a shared target of assembling the full human individual out of different parts. They are reminiscent of Rig Veda's myth about Purusha, the first man-cosmos, who was torn into different parts that became separate cosmic realities, like people, gods, plants, animals, and stars. Now, various humanistic approaches try to do something opposite – to 'reassemble the primary Purusha', i.e. to build an integral human personality into full potential. Someone might criticize it as yet another Tower of Babylon, but its proud view of human nature seems justified to me. It claims that deep down all people are all right, and all our instincts are not animal-like but already human, thus they do not need to be "tamed" or suppressed to make us cultural beings, and we just need an adequate realization of the creative, constructive potential of instincts.

Trans-spirituality

Some transhumanists have a deeper approach which includes the spiritual dimension of the human project – like Buddhist James Hughes, V. R. Manoj, a follower of Sri Aurobindo, or Zen Buddhist Michael La Torra. Manoj argues that the limitation of present day mainstream transhumanism is omitting the most fundamental question about the meaning of human existence "Why we are here and where are we going?"[16] It is this question that has to become the cornerstone of human transformation. The approach of trans-

---

16 V.R. Manoj, "Spiritual Transcendence in Transhumanism", in: *Journal of Evolution and Technology* - Vol. 17, Issue 1 (March 2008), pp. 36-44

spirituality tends to distinguish between religion as a method and spirituality as a goal, where the objective of spirituality is achievement of the state of clarity about human existence, which is also called Enlightenment.

James Hughes argues that the project of human enhancement is consistent with some approaches in spiritual metaphysics, soteriology and eschatology, especially with those approaches that stress "human agency and evolution to a transcendent state, such as Buddhism, or that have incorporated Enlightenment values, such as liberal Christianity."[17] This makes it quite logical and natural that trans-spirituality will appear as a result of the adoption of human enhancement technologies into the theologies of all the world's faiths. The goals which he sets for the trans-spiritual research, such as achieving a comprehensive understanding of spiritual phenomena, and developing techniques for inducing and controlling these phenomena, are quite to the point from the perspective of the human project. His focus seems to be the examination of the physiological and neurological correlates of spiritual experiences, carefully avoiding reductionism. In fact, in his claim that authentic spirituality and deeper science will ultimately coincide, Hughes seems to approach the integral perspective.

Manoj is willing to use science, reason and technology as a means to achieving such a state and to overcome all possible limitations of human existence. At this point the spiritual quest we discussed above and trans-spirituality are in agreement. Manoj receives his inspiration from Aurobindo, who suggested that humankind shall evolve and bring the spiritual reality down to the human level and express it in everyday life. This can also be viewed as a manifestation in the human being of the infinite potential of our transhuman development. Manoj takes this dream seriously and is willing to apply technology for its realization. He claims that religion has its foundations in spirituality which stresses the need to realize superhuman potential, referring for instance to the poem "The Progress of Man" about the evolution of the soul by Sufi philosopher

---

[17] James Hughes (2007), "The Compatibility of Religious and Transhumanist Views of Metaphysics, Suffering, Virtue and Transcendence in an Enhanced Future", in: *Global Spiral*, 8 (2).

Rumi.[18] The important message of Manoj is that stereotyped conflict between science and religion is actually missing the point. He insists that both science and religion are important ways for realizing true human nature in transhumanism. He sees congruence between authentic spirituality and striving to use all possible technical and scientific means to overcome human limitation: "Although I am a deeply religious person, *I do not believe that God wishes us to have limitations.*" [Italics added– T. K.][19] The Sacred as that which inspires humans to transcend their limits, to become free, to dare and create – this idea answers to religious objections to transhumanism which we will discuss now.

Religious objections to transhumanism

Here I will discuss some religious criticisms directed against the link between transhumanism and spirituality. To sum up, it is claimed that transhumanist spirituality is impossible because transhumanism does not respect human rights. Firstly, the association between transhumanism and Nazism is sketched; secondly, an attempt is made to discard the entire need to change the human condition on the grounds that God gave it; thirdly, it is claimed that the present human condition is good enough for moral and spiritual improvement and does not need to be changed. Let us look closer at these objections.

In this chapter I have referred to Nietzsche's ideas on the superhuman. These have led to critique concerning the misuse of his philosophy by Hitler. In my view, the fact that Nietzsche's ideas were misused by Hitler (*Herrenmenschen*) should not make the philosopher accountable. Nietzsche was seeking spiritual renewal, while Hitler wanted what he saw as the biological improvement of the German race; the difference between *Übermensch* and *Herrenmensch* is thus insurmountable. As Solov'ëv has argued, Nietzsche's *Übermensch* is a limited representation of the Divine-Human idea. Even though the idea of *Übermensch* has significant faults (as elitism and egoism, rejection of common human ethics), these do not discard its reminder about human potential. Because of this, Sovlov'ëv, despite

---

[18] Ibidem

[19] Ibidem

all his critique of Nietzsche, at the end accepted him as a co-searcher for spiritual truth.

Indeed, established religions tend to discard the entire quest for transgressing the human condition because it springs from rather different metaphysical foundations. The major difference between philosophers like Solov'ëv or Teilhard de Chardin, on the one hand, and established theology, on the other hand, is their vision of the relation between man and God. For traditional religion, man is a creature whose condition is given by God and must be accepted as it is. Salvation is not something that humans must take responsibility for, the human role is to pray and hope that salvation comes from above. But this leaves so little possibility for one's own initiative that it hardly can satisfy contemporary people.

A different view is offered by spiritualitist humanists like Solov'ëv or Teilhard de Chardin, or Sri Aurobindo. Their basic metaphysical postulate claims that humans have an inborn Divine core and can become active agents of their own destiny, applying their own forces in order to be saved or to achieve higher self-realization. Humans share creativity with God, and human history is a common Divine-Human deed. Spiritualist humanists share this rebellious stance with the Gnostics and with Meister Eckhart, who was almost condemned for claiming that there is uncreated light within the human being. Indeed, if the uncreated Divine nature exists within each human being, then we *must* dare to alter the human condition, and it is all right to use genetic enhancement or any other technology to take control of human evolution. This standpoint falls within the framework of *panentheism*[20] but not pantheism, as it counts on the cooperation with the transcendent Absolute.

This position allows for the conscious control of human biological evolution, *as far as it serves the highest good of all humanity and is done with ethical means.* This view of unorthodox spiritualists on the human being differs from the views of traditional religions which will not agree with man taking an active role in controlling evolution or transgressing the human condition: traditional religion considers humans to be creatures, not inspired and proud co-creators.

---

[20] Panentheism is the view that the universe is embraced by the transcendent Absolute.

Another objection is that unlimited longevity is not necessary for becoming wise, moral or spiritual. Indeed, there were humans who died young but accomplished a lot, achieving outstanding spiritual development. Yet is it possible to achieve "enough" of wisdom or spirituality? Which amount of wisdom is enough, if we take seriously the call "to be perfect as your Heavenly Father"? If the real goal is not simply to achieve some wisdom or goodness, but become perfect, anything less than infinity will fall short. And isn't it infinity that always inspired humans – be it the infinity of the starry sky outside us or of the ethical law inside us, to rephrase Kant? He also called the unlimited life as one of the conditions for finding meaning. As Hermann Hesse wrote in his *Steppenwolf*, spirituality needs lots and lots of time, because to be spiritual one has to feel free from time. Isn't it Infinity as such that symbolizes the Spirit? Or do the proponents of traditional religion find that going beyond the limits of the current human condition is insulting to Jehovah, who jealously guards the prohibited fruit of eternal life and knowledge? Whoever it might be, the one who wishes humans to have limitations, cannot truly be God.

Also connected with the idea that the human condition should not be tampered with (perhaps because it is God-given) is the criticism that transcendence achieved by altering the limits of the human condition is not worth it (akin to 'doping', as one critic put it, and also detrimental to the autonomy of the subject: growth which is not earned). However, there is a difference here, which critics fail to notice, between physical enhancement and spiritual growth. Real spiritual or personal growth simply cannot be the result of enhancement. Technical improvements or genetic enhancement can influence only the external conditions of human existence. They can be facilitators, freeing more time and energy for inner quests via extended longevity, increased memory, enhanced immunity and energy levels. Still, these *technological means can never do the job of growing for our spirit: they are means, not causes.* It is really absurd to think of 'automated happiness' or 'wisdom'. It is simply impossible to make spiritual development automated or technologically enhance it. However, there is nothing wrong with giving people more time or energy, or concentration power to accomplish what they intend. After all, who among us has never worried about deadlines? Deadlines are external, while spiritual efforts are internal by their nature and will always remain so. They can be neither faked nor enhanced.

Physical enhancement is merely a better environment, not an equivalent for psycho-spiritual development, personality integration and meaningful life. Even though some transhumanists indeed confuse these two things, the spiritual tendency within transhumanism seems to overcome this confusion, as we can see from the works of James Hughes or Manoi. One has also to remember the risks of transhumanism, because technology can be detrimental to spirituality and to human nature, if applied in a wrong way. One may even ask: is it good to play with human biology, if the risks are so high? The need for such enhancement is also high, and if it is done in an extremely precautious way, maybe the risks can be minimized.

In my view, the two sides of the enhanced body and infinite intentionality shall complement each other: *mens infinita in corpore immortale.* This complementing illustrates the potential for cooperation between transhumanism and the spiritual quest. Of all transhumanist goals (including e.g. space travel) unlimited longevity is the clearest meeting point. There is nothing absurd in transhumanism. Being an external complement to the inner path to perfection, it can be of enormous help, if applied wisely. Transhumanism can be part of a cooperative and potentially fruitful effort. Although transhumanists and spiritual quest supporters climb the mountain from different sides, it is clear that they belong to the same alpinist team.

Literature

Bailey, Ronald (2004). "Transhumanism: The Most Dangerous Idea? Why striving to be more than human is human", at http://www.reason.com/rb/rb082504.shtml

Bostrom, Nick (2003). "Human Genetic Enhancements: A Transhumanist Perspective", at www.nickbostrom.com/ethics/genetic.html

Dublin, Max (1992). *Futurehype: The Tyranny of Prophecy*, New York: Penguin Books

Eliade, Mircea (1999). *Yoga. Immortality and Freedom* (in Russian), St. Petersburg: Lan'

Fukuyama, Francis (2004). "Transhumanism", *Foreign Policy*, September/October

Hughes, James (2007). "The Compatibility of Religious and Transhumanist Views of Metaphysics, Suffering, Virtue and Transcendence in an Enhanced Future", in: *Global Spiral*, 8 (2)

Manoj, V. R. (2008). "Spiritual Transcendence in Transhumanism", *Journal of Evolution and Technology* - Vol. 17 Issue 1 – March 2008 – pp.36-44

McKibben, Bill (2001). *The End of Nature*, Gloucester MA: Peter Smith Pub.

McKibben, Bill (2003). *Enough: Staying Human in an Engineered Age*, Los Angeles: Times Books

More, Max (1994). "On Becoming Posthuman", at http://www.maxmore.com/becoming.htm

Solov'ëv, V.S., "Smysl žizni" (The Meaning of Love), in: *Sobranie Sočinenij* (Collected Works), Brussels, V. 7., p.31

Solov'ëv, Vladimir (1995). *Lecturers on Divine Humanity*; tr. B. Jakim, Hudson, New York: Lindisfarne Press

Walker, Leslie (2004). "Kurzweil's Quest For Eternal Youth Sets Group Abuzz", at http://www.washingtonpost.com/wp-dyn/articles/A11564-2004Oct6_2.html

Wilber, Ken. Four quadrants scheme: http://wilber.shambhala.com/html/books/cowokev8_intro.cfm/xid,5431661/yid,21177236

9

Worldviews, Genomics and Enhanced Human Lifespan

Peter Derkx

Especially since 1995 there has been a striking increase in the number of research projects on aging featuring genetics. In connection with this development research has moved from description of specific age-associated diseases to explanation and even prevention of underlying aging processes. In this chapter first the question is tackled what 'substantial extension of human lifespan' would mean, and whether this would be a real possibility. The main aim of this contribution, however, is to answer the ethical question whether we should want to live healthy human lives that are decades longer than they usually are now. Will this improve our well-being? Will it benefit everybody equally? Does it foster a meaningful existence? Should we spend large amounts of money to achieve it?

Life-extension and genomics

Enhanced, in the sense of substantially extended, human lifespan has recently become a subject of lively debate. One reason for this is the completion of the Human Genome Project in 2001. Senescence and longevity are influenced by genes. This manifests itself in a number of ways. First, the lifespans of human monozygotic twin pairs are statistically more similar than lifespans of dizygotic twins. Second, there are significant differences in lifespan between different genetically inbred strains of any given laboratory animal, such as the mouse. Third, studies of simple organisms like fruit flies, roundworms and yeast have identified gene mutations that affect duration of life. However, although genes influence longevity, studies of twins and long-lived families have shown that genes account for only about 25% of the variance in human lifespan.[1]

Practically all biogerontologists agree nowadays that humans are not genetically programmed to age and die. We are programmed to survive and reproduce, but natural selection for that has its side-effects, one of which is senescence. Evolution did not select for mechanisms for very long survival of adults after they have reproduced. The human body has an impressive array of maintenance and repair mechanisms, but there has not been sufficient selection to make them perfect. Senescence (physical aging) 'is nothing more nor less than the gradual, lifelong accumulation of subtle faults in the cells and organs of the body'.[2] In this way the mortality risk increases over the life course. There are no genes that cause (normal) senescence, but natural selection did play its part in shaping the process of human senescence. Genes regulate the effort our bodies invest in maintenance and repair and in the way they do this many evolutionary trade-offs are involved, in general between somatic maintenance and repair on the one hand and early growth and reproduction on the other. 'For example, genes that ensure a powerful immune response to infection promote early life survival, but later contribute to inflammation, a major age-related phenotype and risk for developing many diseases'.[3] Moreover, a second specific trade-off seems to be connected with this first one: women with an innate, pro-inflammatory immune system are better protected against fatal infectious disease but are likely to have greater difficulty conceiving a child.[4]

Apart from the 20 to 30% of the genetic factor at least four other major elements contribute to the individuality of the process of human physical aging: type of nutrition, lifestyle (e.g. exercise, excessive food, social isolation), environment (e.g. education, housing and income) and chance. These five elements interact with each other. Nutrition e.g. can influence gene expression. 'Improvements in sanitation, education, nutrition and medicine afforded by increased

---

[1] Westendorp & Kirkwood, 2007, p. 17; Kuningas et al 2008, p. 270

[2] Westendorp & Kirkwood, 2007, p. 20

[3] Vijg & Campisi 2008, p. 1068

[4] Westendorp & Kirkwood 2007, p. 34

wealth are positive indicators for long-term health and decreased early mortality for entire populations'.[5]

What promise does genomics hold for the future of life-extension? The present-day legitimacy of biogerontology as a scientific discipline might be said to derive in large part from its connection with genetics and genomics. The website of the American Federation for Aging Research in 2009 proclaims 'There is no fountain of youth. But there may be genes'.[6] Between 1975 and 2004 there was a steady growth in the number of research projects on aging funded by the American federal government and featuring genetics. The increase has been especially striking since 1995, from 1431 to 2003 projects, an increase of 40%.[7] Before the genomic era research on physical aging was mostly describing the manifestations and effects of age-associated diseases, now it is looking for the root causes of senescence, aiming at explanation, intervention and prevention.

Biogerontological research involving genomics: what type of research is it? Can results be shown already? Researchers from Leiden University recently listed 14 genes which mostly since 1995 have been shown to influence lifespan in yeast, roundworms, fruit flies and mice and which have human homologues. 'In some cases, mean and maximum lifespan is extended up to fivefold. The equivalent life-extending effect in humans would result in an average lifespan of 400 years and a maximum lifespan of over 600 years'.[8] The functions of the key 'longevity' genes and molecular pathways that have been identified so far are: regulation of metabolism, insulin signaling, regulation of growth and development, stress resistance, antioxidant activity, DNA repair, protein repair, tissue proliferation and tumor suppression. I will describe one example to give some idea of what this research is like. In the early 1980s Michael Klass started a search for long-lived mutant roundworms. He discovered that a mutational defect in a single gene, later named age-1, could extend a worm's lifespan. A decade later Cynthia Kenyon discovered a mutant worm which had a life span of 40 days. All mutants of this kind lived

---

[5] Westendorp & Kirkwood 2007, p. 0

[6] http://www.afar.org/

[7] Fishman et al 2008, pp. 300-301

[8] Kuningas et al 2008, p. 274

twice the normal lifespan of 20 days. The cause of the longevity was a defect in a gene called daf-2, which makes a protein similar to the mammalian receptors for two hormones, insulin and insulin-like growth factor 1 (IGF-1). These hormones control the way that animals respond to food, in their metabolism and in their growth. Normally, roundworm insulin-like hormones speed up senescence, but if daf-2 is knocked out, worms live longer, actively and fertile. Kenyon also identified daf-16, a gene that promotes youthful vitality. This gene is turned off by the insulin/IGF-1 signaling (IIS) pathway, and if daf-2 or age-1 are turned down, daf-16 becomes active and extends lifespan. By 2004, through combined manipulations of genes and environment as well as microsurgery, both the Kenyon team and researchers in Belgium had achieved up to six-fold increases in roundworm lifespan. The IIS pathway seems to have evolved very early in the history of life and it seems to have been conserved in worms, fruit flies, mice and also humans. Linda Partridge discovered that the loss of a single copy of chico, a fruit fly gene in this pathway, results in flies showing an extension of lifespan of around 40 per cent, although it also causes dwarfism.[9] In vertebrates like mice the IIS pathway is more complex and contains separate receptors for insulin and IGF-1, but experiments with mouse models by Andrzej Bartke and others show that reduced IIS can extend lifespan also in mammals. The data about humans, gained from studying centenarians and patients with defects in the IIS pathway, indicate that insulin responsiveness influences longevity, that many gene variants are involved here in interaction and that the resulting lifespan extension is much smaller than in the case of roundworms. This is only one example of the research taking place, but the Leiden University researchers conclude:

> In general, the accumulated data have revealed that when moving up the evolutionary ladder, together with an increase of genome complexity, the impact of candidate genes on lifespan becomes smaller. The presence of genetic networks makes it more likely to expect impact of variation in several interacting genes to affect lifespan in humans.[10]

---

[9] Melton 2006, pp. 22-24

[10] Kuningas et al 2008, p. 270

So average and maximum human lifespans of 400 and 600 years respectively, extrapolated from experiments with roundworms, are not to be expected very soon. More on possibilities and expectations in the next section.

Substantial extension of human lifespan: what are we talking about?

Before embarking on a normative discussion it has to be clear what we mean by 'substantial extension of human lifespan'. We can distinguish between four possible outcomes of a biotechnological enhancement of the human lifespan. Varying on work by Harry Moody and Eric Juengst and others we can term these extended morbidity, compressed morbidity, decelerated senescence and arrested senescence.[11]

Extended morbidity means that the average human life becomes longer because the period of (co)morbidity at the end is lengthened. Through good hygiene, nutrition, education, housing, medical care, welfare arrangements and social services old people with one or more chronic diseases stay alive longer. This means that average life-expectancy increases, but this need not be an increase in human flourishing and cause for joy. Extended or prolonged morbidity does not imply an increase in maximum human life-expectancy. A typical time-structure for a human life with extended morbidity could be: growing up from 0 to 20, adult health-span 20 to 55, period of growing morbidity up to 95 as the average age at death and with an unchanged maximum of around 120. Extended morbidity is a scenario some scientists fear as the most likely one, with Alzheimer's disease as one of the main threats, especially for women.[12] Thus Paul Baltes wrote about the good health in the Third Age between 60 and 80 as the good news and about the massive suffering in the Fourth

---

[11] Moody 1995; Juengst et al. 2003, pp. 24-28

[12] Women have a higher risk for Alzheimer's than men, age for age, and women reach higher ages: by age 85 women outnumber men 2 to 1, although starting out life in approximately equal numbers. Moreover, Alzheimer's does not just affect women in greater numbers because they are more likely to get it. They are also more likely to end up taking care of someone who has it, like a spouse, a parent, or a sibling (Legato 2005, pp. 212-213).

Age after 80 or 85 as the bad news.[13] Since nobody wishes this prospect to become reality, we shall not discuss the ethical desirability of this type of life-extension here.

Compressed morbidity is a scenario in which the onset of serious age-associated maladies is delayed as long as possible and thus these are compressed into a shorter period. The maximum human lifespan of around 120 is viewed as fixed. The focus of compressed morbidity is that the average human health-span is extended to a much longer period from 20 up to 'the ideal average life span, approximately 85 years'[14] followed by a relatively short period of decline before death, a period of one or two years at the most. The feasibility of compression of morbidity for the life stage between 55 and 85 was first argued for by James Fries in 1980 and it has been embraced by many. Not long ago three officials of the World Health Organization (WHO) wrote that Fries's tenets and vision 'now lie at the heart of today's approach to NCDs [noncommunicable diseases], ageing and health with its focus on the life course, health promotion, and "active ageing" [use it or lose it]'.[15] Because its original assumption is that the maximum human lifespan is biologically predetermined at around 120 and that death at an average age of 85 is 'natural' and even 'ideal', it might be argued that compression of morbidity is not a form of substantial life-extension. It has to be noted, however, that several gerontologists are of the opinion that compression of morbidity does not occur and actually is impossible. They think it is highly unlikely that we will be able to increase health-span without simultaneously increasing lifespan and the period of morbidity at the end.[16] What might be possible is not compression of morbidity but of disability. To some extent medicine, social services and adapted environments might be able to suppress or keep in check the disability which is caused by chronic diseases (comorbidity) during the latter part of life.[17] Compression of morbidity as a policy in practice seems to

---

[13] Baltes 2003, p. 17; see also Smith et al. 1999

[14] Fries 1980, p. 130

[15] Kalache et al. 2002, p. 243

[16] Neugarten 1996, p. 343; Westendorp & Kirkwood 2007, p. 31

[17] Minicuci et al 2004; Crimmins & Beltrán-Sánchez 2011. Researchers often speak of disability if you cannot do one or more of the following things: wash yourself, put your clothing on, eat, get in or out of bed, go sit in a chair or get up, go

result in compression of the disabling consequences of diseases and not in lives which are disease-free till just before the end.

Decelerated senescence is the outcome in which the processes of biological aging are slowed down, resulting in a higher average life-expectancy and probably a higher maximum life-expectancy. Decelerated senescence means that the period of good health in a human life is extended (as in the scenario of compressed morbidity), but the period of morbidity remains the same or is lengthened as well (as in extended morbidity). Present developments in industrialized countries are in line with this scenario of retarded, delayed or decelerated senescence, with a rise in average life-expectancy of around 2 years every decade.[18] Richard Miller, a respected biogerontologist, thinks that pharmaceutical and genetic interventions will increase the speed of this type of life-extension. 'Nature can slow down aging, and so, it turns out, can we. There are so far two approaches that work for sure: diminished total caloric intake and changes in genes that regulate the rate of early-life growth'.[19] He expects the following average pattern of a human life to be possible: growing up 0-20, adult health-span 20-90 and period of decline after that with death at an age of 112. Maximum life-expectancy at birth might be 140 years. A recent and clear manifestation of the idea of decelerated senescence can also be found in an article by Jay Olshansky, Daniel Perry, Richard Miller and Robert Butler. They can be regarded as representatives of a growing chorus of scientists calling themselves 'moderate', 'modest' and 'realistic'. They firmly believe that an investment now of 3 billion US dollars annually will make it possible to decelerate aging and the onset of aging-related diseases and disorders among the baby boom cohorts by seven years.

> People who reach the age of 50 in the future would have the health profile and disease risk of today's 43-year-old; those aged 60 would resemble current 53-year-olds, and so on. Equally important, once achieved, this seven-year delay would yield equal health and longevity benefits for

---

to the toilet, prepare a meal, shop for groceries, take care of your finances, make a telephone call or do light housekeeping work.

[18] Oeppen & Vaupel, 2002; Westendorp & Kirkwood, 2007, pp. 27-30

[19] Miller, 2004, p. 233

> all subsequent generations, much the same way children born in most nations today benefit from the discovery and development of immunizations.[20]

Arrested senescence refers to relatively complete control of the biological processes of senescence. In this scenario, aging in the sense of senescence or physical and mental deterioration does not occur anymore, or the human organism is cared for very well (maintenance) and what senescence occurs is periodically repaired by a *rejuvenation* cure. For decades or centuries the chance to die does not increase with age any more, but stays rather constant. People still die, but they no longer die from the slow accumulation of damage and chronic deterioration. Instead they die from accident, murder or war. In this scenario people can become very old. Average life-expectancies of 150, 500 or even 5000 years are thought to be possible.

Talking about engineering arrested senescence may sound as if we have entered the field of quackery, pseudoscience or science fiction. But a number of influential organizations exist which promote anti-aging and eventually arrested senescence: the American Academy of Anti-Aging Medicine (A4M), the Gerontology Research Group, The Longevity Meme, the Immortality Institute, the Maximum Life Foundation, the Life Extension Foundation, the Methuselah Foundation and the World Transhumanist Association. Individuals promoting arrested senescence include Deepak Chopra, Ronald Klatz, Michael Brickey, Jean Carper, Gary Null, Walter Pierpaoli, Johannes Huber, Joao Pedro de Magalhaes, Max More, Nick Bostrom, James Hughes, Robert A. Freitas Jr., Colin Farrelly and Ray Kurzweil.

One of the strongest defenders of the scientific credibility of Strategies for Engineering Negligible Senescence (SENS) is Aubrey de Grey, the Chairman and Chief Science Officer of The Methuselah Foundation (Virginia, USA). He not only forcefully argues that humanity needs to set aside massive sums of money for a War on Aging. Together with relevant specialists he has also outlined and embarked on detailing a set of biotechnological measures we could use to beat the 'seven deadly things' that accumulate with age as side-

---

[20] Olshansky et al. 2006, p. 32

effects of metabolism.[21] He expects that in the period between 2025 and 2040 we will be able to fix these problems (to a large extent through genetic interventions and stem cell therapies) and that around 2050 'robust human rejuvenation' will be generally accessible. He realizes that the first fixes will not be perfect, but they will give us time to develop better repair methods. According to de Grey, cancer is the hardest problem to solve, but he thinks it is possible.

Many respected biogerontologists have attacked de Grey's ideas forcefully.[22] It is important to note, however, that the difference of opinion is often mainly political, ethical and related to funding and especially to estimates about the speed of future developments, not about the possibility of substantial life-extension in itself.[23] Two top biogerontologists, Jan Vijg and Judith Campisi, emphatically state that at this moment it is impossible to know whether in the future 'increasingly sophisticated interventions will negate most or all ageing phenotypes' and whether scientists will be able to design interventions 'to alter the orchestrated networks of cell-cell interaction to increase lifespan'.[24] They have a number of reasons for skepticism about the abrogation of human senescence. A central one is the 'evidence that ageing entails a gradual drift towards more random patterns of gene expression, which might cause organ/tissue failure that cannot be undone by pharmacological or biological intervention' (ibidem). What we can aim at to repair is the different responses to severe accumulated damage: either cancer (uncontrolled cell division) or the result of the cellular processes that defend us against cancer: cell elimination (apoptosis) and cell senescence (the cell permanently halts division and proliferation). Apoptosis and cell

---

[21] de Grey et al. 2002; de Grey, 2003 and 2005; and de Grey & Rae, 2007. The seven categories of damage to be solved are: 1. cell death without matching replacement (especially important in the heart and the brain), 2. unwanted cells, e.g. visceral fat and senescent cells (important in arthrosis and diabetes), 3. nuclear (epi)mutations causing cancer, 4. mitochondrial mutations, 5. extracellular protein/protein cross-links (e.g. leading to high blood pressure), 6. extracellular aggregates (e.g. resulting in amyloid involved in Alzheimer's disease), and 7. intracellular aggregates (e.g. resulting in hardening of the arteries). De Grey opts for repairing damage and not preventing it, because human metabolism seems much too complicated for that, especially if one wants useful results fast.

[22] Warner et al 2005 ; Estep et al 2006 ; Holliday, 2007, p. 74

[23] Derkx, 2008

[24] Vijg & Campisi 2008, p. 1070

senescence 'may gradually cause tissue atrophy, and therefore loss of organ function and regenerative capacity' (ibidem). So here we have another evolutionary trade-off, between cancer and atrophy. Through stem cell transplantation it seems possible, in principle, to counter the adverse effects of cell senescence or apoptosis.

> When damage levels are not high enough to elicit an apoptotic or senescence response, a potentially more serious situation ensues that is difficult to counter: the gradual accumulation of random changes in DNA or protein, turning tissues into cellular mosaics. Such stochastic changes can reset gene regulatory loops, and randomly alter gene expression patterns on a cell to cell basis. These changes, in aggregate, might compromise tissue function, without eliciting immediate cellular responses. Stochastic gene regulatory drift would be difficult to correct and could even occur in stem cells in vivo or ex vivo during expansion for transplantation therapy. (ibidem).

The correction of the many small random changes falls outside the realm of current or foreseeable technology 'because there would be no discernible targets for possible interventions'.[25] Perhaps here we have 'intrinsic' or 'normal' aging to be distinguished from 'pathological' aging and age-associated diseases. And it 'is not clear whether successful intervention in overt disease will ameliorate intrinsic ageing, and thus significantly extend human lifespan'.[26]

Worldviews and life-extension

This chapter is not about the possibility of substantial life-extension, but about its desirability. To get some insight into the way people with different worldviews look at life-extension, I shall start with the major ethical arguments for and against effective substantial extension of human lifespan, even though intricate examination of each will not be possible. For all major worldviews the ethical aspect is important.

---

[25] Campisi, personal communication, UCLA, June 28, 2008

[26] Vijg & Campisi 2008, p. 1068

Henceforth by substantial life-extension I mean decelerated senescence and arrested senescence as outlined above. Decelerated senescence is much more probable as the scenario for decades to come, but arrested senescence is certainly interesting, and cannot be completely ruled out for the long run. From an ethical point of view arrested senescence is significant because it forces us to think in new ways about what we think most important in our lives and societies. This is important even if arrested senescence itself turns out to be impossible.

Interpreting and adding to the framework presented by Stephen Post (2004), I have organized the arguments in the following categories: autonomy, beneficence (including non-maleficence), distributive justice and meanings of life.

Autonomy

It is true that the principle or value of autonomy does not figure prominently in the ethical debate about life-extension. Moreover, it is possible to regard autonomy or freedom not as an overriding principle but as one of the good things we value and therefore discuss it under the category of beneficence. However, various kinds of consent and refusal are crucial in traditional biomedical ethics,[27] so it would seem anomalous not to deal with it explicitly; besides, 'autonomy' is often referred to as a fundamental value in liberal democracies.

The first thing to remark is that substantial extension of lifespan in the sense of decelerated or arrested aging at this moment is (still) very much a collective matter. Decisions to try to extend human life substantially for the time being will very much be political or corporate decisions, for instance about priorities in biomedical research. Thus taking autonomy seriously comes down to taking liberal democracy and citizenship seriously. Of course, there is debate about the best way to shape a democracy in a highly multicultural, globalized and technological era.[28] The democracy debate is not

[27] Beauchamp & Childress, 2001

[28] Habermas, 1998; Benhabib, 1996; Carens, 2000; Korten, 2000 and Marres, 2005

uniquely concerned with matters of lifespan extension, but it is highly relevant for how political decisions on life-extension can best be made.

Decisions about priorities in medical and biotechnological research will have impacts on future generations. We must inevitably act paternalistically towards future generations. We cannot ask them. We can only imaginatively try to judge whether it will be good for them and not harm their interests. In some cases we can take precautionary measures to make it easier for people after us to reverse decisions we have made. This last point implies that we should be cautious with lifespan extension involving manipulation of the germline, so with hereditary genetic engineering. We have to be conscientious in our judgments about what is good for future people. We can say that engineering substantial life-extension gives people a real choice they do not have now. In this view, there might be no problem at all, because people with extended lifespans can always decide for themselves and opt for suicide.[29] But would it not be problematic to create a society in which more and more people require this extraordinary exit strategy?

Beneficence

If society has chosen to allow it and to promote it and individuals have the opportunity to make their own autonomous decisions about life-extension, why should they choose for a much longer life or not? An important aspect of the answer concerns the positive or negative effects of the decision taken; the effects which may reasonably be expected. Beneficence is about the moral duty to contribute to the experience of things human beings value. This includes the duty to prevent harm, that is the experience of things that we value negatively. Now, what this means depends on what we ultimately value positively and negatively. Ultimate positive values that are often mentioned are preservation of (the quality of) life; enjoyment, pleasure, happiness, fulfilment or welfare; personal relationships, human company, friendship or community; self-respect and being respected by others; achievement despite obstacles; self-realization or

[29] Horrobin, 2005, p. 19

authenticity; creation and contemplation of beautiful things; knowledge; freedom, autonomy, independence or power; a healthy mind in a healthy body. Things that are valued negatively are the lack or loss of these things and pain and suffering.

The preservation and continuation of life is often mentioned first, but on closer inspection it is not life itself that we ultimately want, but a certain quality of life characterized by a healthy body and an active mind endowed with memory and not overcome with pain.[30] Christine Overall thinks that the main argument in favor of life-extension is not that life in itself is valuable, but 'that *a longer life is the prerequisite for almost everything else that one might want*'.[31] Because of the many possible things people might judge to be ultimate values, and because in a democracy we want to let people decide for themselves about this as much as possible, Overall's argument looks strong, because it is so general. But a much longer life comes at a price. Jeanne Calment, who died when she was 122, stood at the graves of her husband, all her friends, her only daughter and even the grave of her only grandchild.[32] The vulnerability of human beings is not only physical. It also has a crucial social dimension. Living very long might be only attractive if the people you love also live very long. Moreover, life-extension also comes at a cost literally. Money spent on research into fundamental processes of biological aging cannot be spent on asthma research or on education or art subsidies. And a society with a growing number of elderly and really old persons will need to be reorganized in many ways.[33] Costs and benefits (negative and positive effects) of life-extension and longevity have to be weighed against each other. Because it is not life in itself that we want, there might be things that are more important than a longer life.

The quasi-neutral aspect of longevity – everybody can decide for herself what to do with a longer life – makes it fit nicely into a liberal democracy and an individualistic society. But longer lives can also cause problems for what we think valuable. During the last decade we have seen many discussions about the 'graying' of society and the problems this will cause. There is much more than I will mention

---

[30] Brandt, 1959, p. 342; see also Horrobin, 2006

[31] Overall, 2004, p. 287 (emphasis added)

[32] Baars, 2006, p. 199

[33] Baars, 2006, pp. 59-64; Carstensen, 2009, pp. 5-95

here. I will just deal with a few examples, chosen to illustrate the nature and the variety of the problems.

The danger of an overpopulated world and all the problems that go with it has often been put forward as an objection to substantial life-extension. It is important to realize that *overpopulation* is not simply a matter of the number of people, but even more a matter of the ecological footprint of human beings. As Mark Hertsgaard wrote in 1998

> An average American uses 185 gallons of water a day for household tasks; the average Senegalese uses 8. An American consumes about fifty-three times more goods and services than a Chinese. Because similar disparities pertain to many commodities and activities, a baby born in the United States creates thirteen times as much environmental damage over the course of its lifetime as a baby born in Brazil, and thirty-five times as much as an Indian baby.[34]

The relative numbers are changing quickly, especially for China, India and Brazil, but it remains important to make this point. It is also important to realize that population growth is not just the result of a lower death rate. The birth rate is at least as important a factor and a general phenomenon is that birth rates tend to fall as life-expectancy increases.[35] Reproductive decision-making governing fertility and parental investment might be driven by a human psychology designed by natural selection to maximize material wealth, not just the largest number of surviving and reproducing children.[36] Many other, highly unpredictable factors are in play. To mention just one more: 'human vulnerability [might increase] due to new infectious diseases or antibiotic resistant strains of bacteria. (…) Disease may well continue to be an effective leveler, improving its technology as we improve ours'.[37] However, if substantial life-extension is achieved and the death rate goes down dramatically, would the decline in birth rate keep pace with it? Or would governments have to resort to drastic measures, prohibiting either life-extension or children? It strikes as a

---

[34] Hertsgaard, 1999, p. 196

[35] Dykstra, 2002, p. 8

[36] Borgerhoff Mulder ,1998, see also Hrdy, 1999

[37] Harris 2003, p. 75

decisive argument at first when de Grey or de Magalhaes writes that it is morally unacceptable to let old people die in order to solve problems of an overpopulated world. However, is it really obvious that extending the lives of people who already exist is always better than choosing for children? Even recognizing the fact that people who do not yet exist cannot suffer, a society with children might still be better than one without for people who do exist.

Another social problem is presented as a beneficence argument in favour of the acceptance of substantial life-extension. Stephen Post argues from his knowledge of Alzheimer's and the pain and suffering and lack of community, self-respect and independence that go with this disease, and he writes:

> The stark reality of our already aging societies is that … [m]any will experience chronic illness for which old age is the dominant risk factor, ranging from Alzheimer's and Parkinson's to osteoporosis and vascular disease. … The solution to this problematic of age-related disease may rest with advances in the basic science of aging that would achieve even greater prolongevity in a manner that avoids the massive debilitation that currently plagues us.[38]

Because we might be able to *prevent* the strong negative value of *age-related chronic diseases* by anti-aging interventions, we must develop these. They will lead to extended life-expectancy, but according to Post this is a necessary price to pay for success in the fight against horrible suffering (if it is a price). Rudi Westendorp, a Dutch biomedical gerontologist, takes a similar view, but he warns that effective therapies for the age-related diseases we know now will extend average life-expectancy and this will unmask new diseases related to the longer lives that will be possible then. Therapies will be needed for newly-discovered diseases we now know hardly anything about.[39] And so on and so forth. The argument by Post turns out to be not so self-evidently in favour of anti-aging interventions. His solution is not a definitive one. Hard choices will have to be made.

Although new technological possibilities, expensive drugs and new diseases like obesity will also be very important factors contributing to the expected rise in medical expenditure, 'an increase

---

[38] Post, 2004, p. 537

[39] Köhler, 2004

in life-expectancy that is not accompanied by a proportionate increase in healthy years will lead to a great *increase in healthcare costs*'.[40] In most industrialized countries the percentage of health care resources spent on the elderly over age 65 is already much greater than the percentage spent on the entire population under age 65.[41] Estimates are that by 2050 the US will spend more than a thousand billion dollars annually only on Alzheimer's disease and related dementias.[42] Will we be able to pay for an expanded scenario along these lines? Will the costs be so high as to hinder all other social priorities and things we value personally? Besides, because wealthy people live longer, schemes like the American Medicare program (309 billion dollars in 2006) may be thought to contain an ethically dubious element. To a significant extent they are equivalent to channelling tax revenue from the population as a whole towards expensive care for the fairly well-off.[43] And as the aging populations of developed countries are coming to require more medical attention and suck away local doctors and nurses from developing countries, shortages of health-care workers in the developing world create ever more desperate situations.[44]

Distributive justice

Ethics is not only about promoting the good and preventing the bad things in life and the real possibility of shaping it yourself, it is also about the distribution of all this. Justice is about the distribution of the (lack of) things we value – such as freedom, happiness, friendship, beautiful things and good health – and the things we do not want – a life solitary, poor, nasty, brutish, and short. At the beginning of this article I have distinguished between all kinds of life-expectancies, but I left out one very important factor: the 'social gradient' of longevity. Life-expectancies differ according to social

---

[40] Knook, 2002, p. 21 (emphasis added)

[41] Beauchamp & Childress 2001, p. 260. I must say that it has not become clear to me what exactly is counted as 'health care resources spent on the elderly'.

[42] Olshansky et al. 2006, p. 31

[43] Wijnbergen, 2002, p. 7

[44] Garrett, 2007, pp. 15 and 26

status. Michael Marmot's summary starts with an illustration from the United States capital:

> If you take the Metro from the southeast of downtown Washington to Montgomery County, Maryland, in the suburbs – a distance of about 14 miles – for each mile traveled life expectancy rises about a year and a half. This is the most life-enhancing journey in the world. There's a twenty-year gap between poor blacks at one end of the journey (male life expectancy fifty-seven), and rich whites at the other.[45]

Such inequalities in life-expectancy at birth are enormous and they are universal. They exist all over the world. Socio-economically determined inequalities in healthy life-expectancy are even longer.[46] What counts as injustice depends on the theory of justice that is used. However, whether one refers to human rights, Rawls' theory of justice as fairness, Dworkin's equality of welfare and resources or Nussbaum's capabilities theory,[47] differences in average life-expectancy at birth of forty years between countries (Japan and Zimbabwe) and twenty years for socio-economic groups within countries – differences which can be removed and prevented by collective social action - are hard to defend as morally acceptable. Now imagine what would happen if in such a world substantial life-extension became possible through initially very expensive biotechnology like longevity pharmaceuticals or gene therapy. The demand backed by purchasing power, certainly in the beginning, would mainly come from young adults, the better educated, wealthier and higher-income individuals and those with higher initial endowments of health. Socioeconomic and health inequalities would be amplified. A small group of people with an already high life-expectancy would have access to life- and health-span-extension, many less-privileged people would not. Surely this is undesirable, is it not? 'The need-based claims of the worse off to have reasonably long

---

[45] Marmot, 2005, p. 5

[46] Marmot, 2004, pp. 13-36; Mackenbach & Bakker, 2003; van der Lucht & Polder, 2010, pp. 53-61

[47] http://www1.umn.edu/humanrts/index.html; Buitenweg 2007; Rawls 1999; Dworkin 2000; Nussbaum, 2011

lives have more moral weight than the preference-based claims of the better off to have longer lives'.[48]

But the existence of social injustice can never normally be a valid reason to object morally to any improvement in the fate of human beings who do not belong to the most underprivileged ones. As Stephen Post writes:

> If we were to insist that technological developments of all sorts wait until the world becomes perfectly just, there would be absolutely no scientific progress. Requiring equality as the prerequisite to biogerontological advance is to establish an obstacle that is virtually insurmountable, and so exceedingly high as to be implausible. Indeed this is not a requirement imposed in any other area of scientific research and development, from new dental treatments to organ transplantation. ... Anti-aging research and, eventually, derived treatments, will emerge in technologically advanced countries and be affordable to those who can pay. This is the unavoidable future of all biotechnological efforts in human enhancement. And yet scientific creativity of this sort will not be inhibited.[49]

Post is right in many ways. Demanding equality and perfect justice within and between countries as a prerequisite to the development of life-extension technology is asking too much. Here as often 'perfection' would be the enemy of the good. On the other hand, not being able to do everything, or enough, should be no excuse for doing nothing. Efforts like the UN Millennium Development Goals[50] are very important. It is important before 2015 to try to reduce by half the proportion of people living on less than a dollar a day, to reduce by two thirds the mortality rate among children under five and to try to reduce by three quarters the maternal mortality ratio.[51] These are challenging goals but they are technically feasible and mainly depend on social responsibility and political will. In the same vein ambitious but feasible goals could be formulated to do something about the shocking disparities in longevity between and also within

---

48 Glannon, 2001, p. 167; see also McConnel & Turner, 2005, p. 61 and Mauron 2005

49 Post, 2004, p. 537; see also Harris, 2003 and Davis, 2004

50 http://www.un.org/millenniumgoals/#

51 Garrett, 2007, p. 32

countries. Within countries Christine Overall proposes a qualified prolongevitism which will genuinely be for all, a kind of affirmative action in the field of life-extension.

> As a general principle, support for increased longevity should not be limited by gender, socioeconomic class, sexual orientation, race, or ability. So the particular focus, at least in the short run, of measures to increase average life expectancy must be on members of groups that historically have been disadvantaged and that currently have low life expectancy.[52]

This seems to imply that increased research into conditions and diseases that affect groups of people with low life-expectancy, like black and poor people, is morally indicated. However, note that for longer life-expectancies of less-privileged people more equality of income is more important than new achievements in high-tech biomedicine. Only three of the thirty-nine recommendations in the Acheson Report to the British government - *Inequalities in Health: Report of an Independent Inquiry* (1998) – related to health care. 'The others covered the tax and benefit system; education; employment; housing and environment; mobility, transportation, and pollution; and nutrition'.[53] Reducing health inequalities might both be the ethically indicated and the most effective way to substantially extend the (remaining) life-expectancies of many people; more effective than biotechnological approaches aimed at decelerating or arresting senescence of human beings as a species. Many healthy human years can be gained by this heavily neglected form of life-extension: **more equal longevity**. This kind of priority-setting might provoke strong political opposition, but that is no reason to be silent about one's considered ethical judgment.

Furthermore, it is important to see that priorities do not have to be of the all or nothing type. Serious and strenuous attempts to tackle the national and global social gradient of longevity certainly do not require that biogerontological research into the diseases of the oldest old and general underlying processes of senescence is stopped completely. As far as international injustice is concerned, one should not forget that the numbers of the old and oldest old in developing

---

[52] Overall, 2003, p. 200

[53] Marmot, 2004, p. 251

countries will also increase rapidly. Already the remaining life-expectancy of a woman who has managed to reach the age of sixty in Brazil (21 more years), India (18 years) and Nigeria (17 years) is not so different from the number of years an average sixty-year-old female inhabitant of the United States can expect to add to her life (24 years). The WHO anticipates that the percentage of people above sixty living in developing countries between now and 2050 will rise from 60 to 85% of the total global number.[54] In China the elderly will outnumber the total current population of the US by mid-century, more than 300 million people. The same goes for India.[55]

Meanings of life and life-extension[56]

So far we have seen that autonomy, beneficence and justice evoke important ethical considerations when thinking about life-extension. Our task seems to be one of weighing pros and cons, in a situation of risk and uncertainty about the future. These ethical considerations very much seem to be important for everybody, whatever her or his worldview, religious or non-religious. However, some influential authors already attempt a more definitive position. They do not want to wait and see how things develop. They take a principled stand against life-extension now. Hans Jonas, Leon Kass, Francis Fukuyama, Daniel Callahan and Bill McKibben represent what one might call a natural law position. Not only is this a significant stance in itself, its discussion points very clearly to the way in which issues of meaning recur in the discussion of life-extension. Important aspects of the natural law position are: that aging is the final stage in a natural life cycle which should be cultivated,[57] that aging is a natural process to be accepted and not a disease to be defeated,[58] that the

---

[54] Kalache et al. 2005, p. 36; see also Aboderin, 2006 and Kirkwood, 1999, p. 8

[55] Olshansky et al. 2006, p. 31

[56] Autonomy, beneficence and justice are also very relevant for the experience of a meaningful life, but the human needs for meaning include more, notably needs for purpose, fulfillment, comprehensibility, coherence, transcendence and feeling one with or part of a valuable whole (see e.g. Smaling & Alma 2010 and Derkx 2011).

[57] Callahan, 1995, but Overall, 2003 and 2004

[58] Callahan, 1995, but also Moody et al. 2004; Caplan, 2005; Izaks & Westendorp, 2003; Scully, 2004, and Derkx, 2006

goals of medicine and health care include therapy (treatment) and prevention but not enhancement,[59] and that it is unnatural and selfish to prefer a society with many very old people and very few children over a society with fewer elderly persons and a more natural succession of 'fresh' generations.[60]

It is very easy to dismiss natural law positions as an untenable deontological stance by pointing out that if substantial life-extension starts to occur in nature it begins to be 'natural', or by emphasizing that humans have always changed nature (including their own natural features) in the course of civilization. More or less the same goes for the religious versions of these arguments, referring to a God who has established the natural law. That humans should respect the will of God or that they should not attempt to play God, runs into the same kind of intellectual difficulties as the exhortation to respect nature, and into additional difficulties as well. Referring to the will of God is not a very strong argument in a pluralistic democratic society which includes atheists and agnostics.

However, it is possible to discover something important behind these arguments from nature or God, even if one rejects the absolute verbal form and is more inclined towards consequentialist ethics. Human nature is not blank nor completely and always easily malleable. It is the result of millions of years of natural selection. Human beings are evolutionary results and as such they are very complex organisms with many trade-offs involved, referring back to environments of the past. We cannot design humans from scratch because we have entered new, highly technological environments. Stressing that we ought to be wary of bad unintended consequences is not the same as claiming that nothing should be changed. It is possible for a society to opt for a less-than-one-child-per-family-policy to counteract undesirable effects of population-growth due to increasing old age survival, but will its individual citizens accept this policy and live up to it? Human nature is very flexible, but it is possible to ask too much of human beings. It seems relevant, for example, to speculate about the emotional implications of a population scenario with 9 billion people in 2300 with an average life-expectancy at birth of about 100 years and a high proportion of very

---

59 Juengst, 1997, 1998 and 2004; Brock, 1998 and Frankford, 1998

60 McKibben in: Stock & McKibben, 2003; Kass, 2004, pp. 317-318

old people.[61] And we should not only be talking here about what is possible for human beings, individually and as a group. We should also be talking about what is good for them and what makes their lives meaningful. To ask what desires and emotions are humanly 'natural' can be translated into questions about the emotional wiring of humans as evolutionary products and questions about what desires and emotions are good and proper for human beings to have and deserve the opportunity to be acted upon.

Authors criticizing substantial life-extension often point to loss of meaning. Thus, Hans Jonas writes: 'Perhaps a non-negotiable limit to our expected time is necessary for each of us as the incentive to number our days and make them count'.[62] And Bill McKibben said:

> Maybe with these tools [such as germline genetic engineering] we will in some way learn to live forever, but the joy of it, the meaning of it, will melt away like ice cream on an August afternoon. … Life far beyond the parameters of what we know now, life that goes beyond the normal human expectations, may be very much like a trap, and the name of that trap is a very American one – the constant idea that more is better. If it is good to live 80 years, it must be better to live 180 years and far better yet to live 300 years.[63]

Yet much can be said in response to these objections to substantial life-extension. What does 'making our days count' exactly mean? Horrobin points out that it is an odd argument to assert that people enjoy playing football today and experience no ennui in doing so only because they are aware that they cannot do it three centuries hence.[64] Moreover, what are 'normal human expectations'? The normal expectations of young Western European female office-workers in the 21st century are very different from those of their counterparts in 1875. And is it natural or unnatural for human beings to think that more is better? Perhaps the most fundamental criticism of the natural law position is expressed by Christine Overall.[65] She states that we

---

61 Basu, 2004, p. 93

62 Quoted in Post, 2004, p. 36

63 Stock & McKibben, 2003, p. 7

64 Horrobin, 2005, p. 14

65 Overall, 2004; see also Overall, 2003

should not argue against increasing human longevity by reference to the limited parameters set by current life-expectancies. According to her this is the fallacy of begging the question. When the context changes and life-expectancies become much longer, our judgment of life's possibilities and meanings will also change. Not only will childhood and age be redefined, but concepts like schooling, education, marriage, partnership, friendship, sexuality, gender, father, mother, parent, grandparent, family, work, job, career, retirement, nationality and citizenship will also take on other meanings. Together these changes will constitute new moral systems. I would agree with Overall, if she argued that we ought not to evaluate substantial life-extension only by reference to the kind of life that we know now. Certainly, a comprehensive ethical judgment about future possibilities requires not just norms, values, facts and extrapolations but also moral imagination. But I think Overall exaggerates. The way we think about human fulfilment now, of course, is relevant to our well-considered present-day judgments on prolongation of life.

A concern with meaning is not the prerogative of natural-law critics of substantial life-extension. Paul van Tongeren does not accept Callahan's arguments in favour of accepting a natural lifecycle and a natural lifespan, but he emphasizes the importance of knowing what the meaning of life is. 'If we ask to what extent we are allowed to, or even should, extend life, we have to realize that we can hardly answer that question as long as we do not know what life is all about.'[66] Whether we think substantial life-extension morally acceptable or desirable very much depends on what we think gives life its meaning. Here Walter Glannon advances an argument from personal identity.

> Would a significantly long span of time between earlier and later mental states weaken the relations between them and make them so different that they effectively belonged to two distinct persons? … Psychological connectedness is necessary for what matters to us.[67]

More discussion about meanings of life is needed. But in individualistic secular societies people have many different ideas

---

[66] van Tongeren, 1995, p. 36

[67] Glannon, 2001, p. 160; for counterarguments see Harris, 2003, pp. 82-85

about meanings of life, so it will be difficult to reach consensus or even understanding about the value of life-extension. Part of the difficulty is that in modern Western societies it is rather generally accepted that meanings of life - and worldviews and religions as meaning frames - are a private matter, not something about which to engage in public debate.

The variety in ideas about meanings of life will be very difficult to handle in a democracy, because the differences can be wide and not a matter of degree. Aubrey de Grey and other transhumanists feel that we should not accept biological aging as inevitable. They argue that the fundamental biology of human beings should be changed in order to get rid of senescence and most of death.[68] Other thinkers, not only of the natural-law variety, see this as a dangerous illusion, holding that the propagation and cultivation of ideas like this are very detrimental to the meanings of human lives. This difference in worldview is an extremely important aspect of the debate on substantial extension of human life-expectancy. What is involved is expressed very clearly in the words of Michael Lerner:

> We need to do the spiritual work as we grow older to accept the inevitability of death rather than acting as though aging and death could be avoided if only we had a better technology. The enormous emotional, spiritual, and financial cost of trying to hang on to life as long as possible (and to look as though we were not aging) is fostered by a marketplace that tries to sell us endless youth. It is also fostered by our cultural failure to honor our elders, provide them with real opportunities to share their wisdom, and combat the pervasive ageism with its willingness to discard people long before their creative juices have dried up, to stigmatize the sexuality of the elderly …, and to provide little in the way of adequately funded and beautifully conceived long-term care facilities.[69]

---

[68] More, 2004; Kurzweil, 2004, Bostrom 2003; Hughes, 2004; Stock, 2002
[69] Lerner, 2006, pp. 308-309; see also Baars, 2006

Worldviews (religious and non-religious) and life-extension

The preceding pages on substantial life-extension started from the presupposition that biogerontology offers many unprecedented promises, but I hope that they also made clear that a gerontology that is exclusively biological would create problems while it tried to solve others. A technological fix for existential problems will not work and might make things worse. Like all other fields of crucial human concern, aging must be studied in an interdisciplinary mood and mode. According to Jan Baars, and he keeps emphasizing this, 'aging' can be studied from different perspectives and acquires different meanings in the process: (1) aging in the socio-political sense of the 'graying' of society: the growth in numbers of persons with increasing chronological age with all its political, social, legal and financial implications, (2) aging in the sense of senescence, biological decline which need not keep pace with chronological age and which biogerontologists would like to be able to measure with 'biomarkers', (3) aging in the existential sense of subjectively experiencing and interpreting one's own unique and vulnerable life as it unfolds in time and society.[70] These different concepts of aging are often intermingled and mixed up, with chaotic and sometimes dehumanizing results. A study of aging which is culturally dominated by the second concept of aging and by the science of biological senescence might turn out to be cruel. Biogerontological research needs to be embedded in a democratic dialogue on meanings, goals and values. In such a dialogue all citizens should be involved and it should not only be fostered by the natural sciences, but also by the social sciences and the humanities. None of these branches of knowledge and research can be practiced without recourse to explicit or implied normative assumptions and human values. Most biogerontologists I know personally would immediately assent to this: they are humanistic scientists (and often know more of philosophy and sociology than humanities scholars know of the natural sciences), but the cultural status and prominence of much-publicized biogerontological progress often has an unintended and undesirable cultural impact.

---

[70] Baars, 2006; see also Baars, 2007

> There is an irredeemable cultural logic – if death is a solvable problem, then old age will be a failure. … Locating the meaning of death in biochemical processes and striving for ever-longer lifespans denies the possibility of old age as a valued final part of the life course.[71]

When I first read this I was inclined to think that Vincent exaggerates. On June 28, 2008, however, I spoke to Aubrey de Grey during the conference 'Understanding Aging' at UCLA and I told him that I head the research group 'Ageing Well' at the University of Humanistic Studies in Utrecht. A bystander thought it an interesting research project, but de Grey said he did not like it: 'The title of the project suggests that aging might be a good thing, which it is not. Aging kills.' This reaction shows the importance of Baars's distinction between the three senses of 'aging'.

Research on meanings of life, the value of old age and a just and humane society requires approaches and methods that fit the subject of investigation, and in this area research which confines itself to a 'hard' laboratory approach is often inappropriate and not the best type of research which is possible. An interdisciplinary humanistic gerontology, however, is a challenging task with risks and dangers on all sides. One important problem can be illustrated by a passage from Clifford Geertz on the religious perspective on suffering:

> The problem of suffering is paradoxically, not how to avoid suffering but how to suffer, how to make of physical pain, personal loss, worldly defeat, or the helpless contemplation of others' agony something bearable, supportable – something, as we say, sufferable….[72]

Suffering is and will be a part of human existence, but the function of religious and other worldviews is not only to learn to accept suffering. Human suffering always has a social, cultural and historical context, which can change. These changes do matter very much and they transform the character of the moral and existential issues we have to deal with. For example: my mother's parents, Roman Catholics living in the Netherlands, had five young children in 1918-1919. Between December 1918 and January 1920 four of them died

---

[71] Vincent, 2006, p. 694

[72] Geertz, 1973, quoted in Cole, 2002, p. 37

of whooping cough, a contagious disease which now can be prevented by vaccination. The small old card commemorating the short lives of the infants starts with the exhortation 'Parents, do not cry!' and ends with the lines 'Say it now and say it forever: what God does, is done well'. Indeed, one important aspect of religious and non-religious worldviews is that they offer a way to accept, endure and embrace human life as it is. A dangerous trap, however, for religious as well as non-religious meaning frames and related ethics is that they become too one-sided. The task for humanists, religious ones included, is to find a wise balance between accepting humanity as it is and aiming for an enhanced humanity that could be. Humanity never just is. Inevitably and in a changing context it is always interpreted by 'human' individuals and in cultures and worldviews.

A tension between progress and acceptance, autonomy and heteronomy,[73] between natality and mortality,[74] between choice and chance,[75] between improving unbearable situations and finding meaning in situations we cannot change (yet): this tension is part of the human condition all worldviews have to deal with. Choosing completely for one of the poles, thereby abolishing the tension, seems to be a strong temptation for human beings, but one we better not succumb to. In a world with ever more elderly persons it is important to continue research into age-associated diseases and the fundamental processes of senescence, in the reasonable expectation that it will alleviate human suffering and promote a better health-span. We do not know whether this research will lead to substantively more life-extension than two years per decade. We do not know whether it will increase maximum lifespan. We do not know whether it will result in cures for Alzheimer's disease, arteriosclerosis and other age-associated diseases. We do not know what 'age-associated' really means and we do not know whether it will become possible to cure or even prevent 'intrinsic aging'. We do know that social inequality leads to glaring inequalities in (healthy) life-expectancy. In such a situation it certainly is important to find out how to age well. In this research it is imperative to distinguish as well as combine the biomedical, socio-political and existential perspectives.

---

[73] Alma, 2005
[74] Wörner, 2000
[75] Dworkin, 2000, ch.13

Acknowledgements

Research for this chapter was supported by a grant (code: 050-32-570) from NWO (Netherlands Organisation for Scientific Research). For some helpful remarks in discussions on the topic of this chapter I should also like to thank Jan Baars, Adri Smaling, Hugo Letiche, Janneke van Mens-Verhulst (University of Humanistic Studies, Utrecht), Dick Knook and Diana van Heemst (Leiden University Medical Center), Pearl Dykstra (Erasmus University Rotterdam), Ricca Edmondson (National University of Ireland, Galway), Robert Binstock, Dale Dannefer and Jennifer Fishman (Case Western Reserve University, Cleveland), Richard Settersten (Oregon State University, Corvallis) and my colleagues in the research project 'Towards a *lingua democratica* for the public debate on genomics', especially Cor van der Weele and Harry Kunneman.

Literature

Aboderin, Isabella (2006). Ageing in Africa. The Wellcome Trust, 1 September 2006 [cited 24 January 2007]. Available from http://www.wellcome.ac.uk/doc_WTX033903.html

Alma, Hans A. (2005). "Humanisme en christendom als bronnen van zin." In *Humanisme en religie: controverses, bruggen, perspectieven*, edited by Joachim Duyndam et al, 339-54. Delft: Eburon.

Baars, Jan (2006). *Het nieuwe ouder worden: paradoxen en perspectieven van leven in de tijd.* Amsterdam: SWP, Humanistics University Press. An English language version of this book will be published by Johns Hopkins University Press.

——— (2006). "A Triple Temporality of Aging: Chronological Measurement, Personal Experience, and Narrative Articulation." In *Aging and Time: Multidisciplinary Perspectives*, edited by Jan Baars & Henk Visser, 15-42. Amityville, New York: Baywood.

Baltes, Paul B. (2003). "Extending Longevity: Dignity Gain - or Dignity Drain?" *MaxPlanckResearch*, 14-19.

Basu, Alaka Malwade (2004). "Towards an Understanding of the Emotions in the Population of 2300." In *World Population to 2300*, edited by United Nations - Department of Economic and Social Affairs - Population Division, 89-98. New York: United Nations.

Beauchamp, Tom L. & James F. Childress (2001). *Principles of Biomedical Ethics.* Fifth ed. Oxford: Oxford University Press.

Benhabib, Seyla, ed. (1996). *Democracy and Difference: Contesting the Boundaries of the Political.* Princeton, NJ: Princeton University Press.

Borgerhoff Mulder, Monique (1998). "The Demographic Transition: Are We Any Closer to an Evolutionary Explanation?" *Tree* 13, no. 7 (July): 266-70.

Bostrom, Nick (2003). "Human Genetic Enhancements: A Transhumanist Perspective." *Journal of Value Inquiry* 37, no. 4: 493-506.

Brandt, Richard B. (1959). *Ethical Theory: The Problems of Normative and Critical Ethics.* Englewood Cliffs, N.J.: Prentice-Hall.

Brock, Dan W. (1998). "Enhancements of Human Function: Some Distinctions for Policymakers." In *Enhancing Human Traits: Ethical and Social Implications*, edited by Erik Parens, 48-69. Washington, D.C.: Georgetown University Press.

Buitenweg, Rob (2007). *Human Rights, Human Plights in a Global Village.* Atlanta, Georgia: Clarity Press.

Callahan, Daniel (1995). "Aging and the Life Cycle: A Moral Norm?" In *A World Growing Old: The Coming Health Care Challenges*, edited by Daniel Callahan et al, 20-27. Washington D. C.: Georgetown University Press.

——— (1995). *Setting Limits: Medical Goals in an Aging Society.* Expanded ed. Washington, D.C.: Georgetown University Press.

Caplan, Arthur L. (2005). "Death as an Unnatural Process: Why Is It Wrong to Seek a Cure for Ageing?" *EMBO Reports* 6, Special Issue (July): 72-75.

Carens, Joseph H. (2000). *Culture, Citizenship, and Community: A Contextual Exploration of Justice as Evenhandedness.* Oxford: Oxford University Press.

Carstensen, Laura L. (2009). *A Long Bright Future: An Action Plan for a Lifetime of Happiness, Health, and Financial Security.* New York, NY: Random House, Broadway Books.

Cole, Thomas R. (2002). "On the Possibilities of Spirituality and Religious Humanism in Gerontology or Reflections of One Aging American Cultural Historian." In *Cultural Gerontology*, edited by Lars Andersson, 25-44. Westport, Connecticut: Auburn House.

Crimmins, Eileen M. & Hiram Beltrán-Sánchez (2011). "Mortality and Morbidity Trends: Is There Compression of Morbidity?". *Journal of Gerontology: Social Sciences*, 66B (1): 75-86.

Davis, John K. (2004). "Collective Suttee: Is It Unjust to Develop Life Extension If It Will Not Be Possible to Provide It to Everyone?" In *Strategies for Engineered Negligible Senescence: Why Genuine Control of Aging May Be Foreseeable*, edited by Aubrey D. N. J. de Grey, 535-41. New York, N.Y.: New York Academy of Sciences.

de Grey, Aubrey D. N. J. (2003). "The Foreseeability of Real Anti-Aging Medicine: Focusing the Debate." *Experimental Gerontology* 38, no. 9 (1 September): 927-34.

——— (2005). "Foreseeable and More Distant Rejuvenation Therapies." In *Aging Interventions and Therapies*, edited by Suresh I. S. Rattan, 379-95. Singapore: World Scientific Publishing.

de Grey, Aubrey D. N. J. et al (2002). "Time to Talk Sens: Critiquing the Immutability of Human Aging." In *Increasing Healthy Life Span: Conventional Measures and Slowing the Innate Aging Process*, edited by Denham Harman, 452-62. New York, N.Y.: New York Academy of Sciences.

de Grey, Aubrey & Michael Rae (2007). *Ending Aging: The Rejuvenation Breakthroughs That Could Reverse Human Aging in Our Lifetime.* New York, N.Y.: St. Martin's Press.

Derkx, Peter (2008). "Onafhankelijkheid: de onafhankelijkheid van de onderzoeker in een multidisciplinaire context: humanisme, ethiek en verouderings-onderzoek." In *Deugden van de humanistiek: kwaliteiten en valkuilen van wetenschap met een menselijk gezicht*, edited by Joachim Duyndam et al, 113-25. Amsterdam: SWP/Humanistics University Press.

——— (2006). "Ouder worden: te aanvaarden natuurlijk proces of te bestrijden ziekte?" *Tijdschrift voor Humanistiek - Journal for Humanistics* 7, no. 28 (december): 82-90.

Derkx, Peter (2011). *Humanisme, zinvol leven en nooit meer 'ouder worden': een levensbeschouwelijke visie op ingrijpende biomedisch-technologische levensverlenging.* Brussel: VUBPress.

Dworkin, Ronald (2000). *Sovereign Virtue: The Theory and Practice of Equality.* Cambridge, Massachusetts: Harvard University Press.

Estep, Preston W., III et al (2006). "Life Extension Pseudoscience and the SENS Plan." *Technology Review* (MIT), July 11 [cited August 15 2006]. Available from http://www.technologyreview.com/sens/index.aspx

Fishman, Jennifer R. et al (2008). "Anti-Aging Science: The Emergence, Maintenance, and Enhancement of a Discipline." *Journal of Aging Studies* 22, no. 4 (December): 295-303.

Frankford, David M. "The Treatment / Enhancement Distinction as an Armament in the Policy Wars." In *Enhancing Human Traits: Ethical and Social Implications*, edited by Erik Parens, 70-94. Washington, D.C.: Georgetown University Press, 1998.

Fries, James F. (1980). "Aging, Natural Death, and the Compression of Morbidity." The *New England Journal of Medicine* 303, July 17: 130-35.

Fukuyama, Francis (2002). *Our Posthuman Future: Consequences of the Biotechnology Revolution.* New York: Farrar, Straus and Giroux.

Garrett, Laurie (2007). "The Challenge of Global Health." *Foreign Affairs* 86, no. 1: 14-38.

Glannon, Walter (2001). *Genes and Future People: Philosophical Issues in Human Genetics.* Boulder, Colorado: Westview Press.

Habermas, Jürgen (1998). *Between Facts and Norms: Contributions to a Discourse Theory of Law and Democracy.* Edited by Thomas McCarthy. Cambridge, Massachusetts: MIT Press.

Harris, John (2003). "Intimations of Immortality: The Ethics and Justice of Life Extending Therapies." In *Current Legal Problems 2002*, edited by M. D. A. Freeman, 65-95. Oxford: Oxford University Press.

Hertsgaard, Mark (2007). *Earth Odyssey: Around the World in Search of Our Environmental Future.* New York: Random House, Broadway Books.

Holliday, Robin (2007). *Aging: The Paradox of Life. Why We Age.* Dordrecht: Springer.

Horrobin, Steven (2005). "The Ethics of Aging Intervention and Life-Extension." In *Aging Interventions and Therapies*, edited by Suresh I. S. Rattan, 1-27. Singapore: World Scientific Publishing.

——— (2006). "The Value of Life and the Value of Life Extension." In *Understanding and Modulating Aging*, edited by Suresh I. S. Rattan et al, 94-105. Boston, Mass.: Blackwell Pub.

Hrdy, Sarah Blaffer (1999). *Mother Nature: A History of Mothers, Infants, and Natural Selection*. New York: Random House, Pantheon Books.

Hughes, James (2004). *Citizen Cyborg: Why Democratic Societies Must Respond to the Redesigned Human of the Future*. Boulder, Colorado: Westview Press.

Izaks, Gerbrand J. & Rudi G. J. Westendorp (2003). "Ill or Just Old? Towards a Conceptual Framework of the Relation between Ageing and Disease." *BMC Geriatrics* 3, no. 7.

Juengst, Eric T. (2004). "Anti-Aging Research and the Limits of Medicine." In *The Fountain of Youth: Cultural, Scientific, and Ethical Perspectives on a Biomedical Goal*, edited by Stephen G. Post & Robert H. Binstock, 321-39. Oxford: Oxford University Press.

——— (1997). "Can Enhancement Be Distinguished from Prevention in Genetic Medicine?" *The Journal of Medicine and Philosophy* 22: 125-42.

——— (1998). "What Does Enhancement Mean?" In *Enhancing Human Traits: Ethical and Social Implications*, edited by Erik Parens, 29-47. Washington, D.C.: Georgetown University Press.

Juengst, Eric T. et al (2003). "Biogerontology, "Anti-Aging Medicine," and the Challenge of Human Enhancement." *Hastings Center Report* 33, no. 4 (July-August): 21-30.

Kalache, Alexandre et al (2002). "Compression of Morbidity and Active Ageing: Key Priorities for Public Health Policy in the 21st Century." *Bulletin of the World Health Organization* 80, no. 3 (March): 243-44.

Kalache, Alexandre et al (2005). "Global Ageing: The Demographic Revolution in All Cultures and Societies." In *The Cambridge Handbook of Age and Ageing*, edited by Malcolm L. Johnson et al, 30-46. Cambridge, UK: Cambridge University Press.

Kass, Leon & Morton Kondracke (2004). Ageless Bodies and Happy Souls: The Future of Aging in a Biotech Era. SAGE Crossroads, April 12 [cited November 15 2004]. Available from http://www.SageCrossroads.net

Kass, Leon R. (2004). "L' Chaim and Its Limits: Why Not Immortality?" In *The Fountain of Youth: Cultural, Scientific, and Ethical Perspectives on a Biomedical Goal*, edited by Stephen G. Post & Robert H. Binstock, 304-20. Oxford: Oxford University Press.

Kirkwood, Tom (1999). *Time of Our Lives: The Science of Human Aging*. Oxford: Oxford University Press.

Knook, D. L. (2002). "Europe's Ageing Population and Consequences for Health Care." In *Ageing in Europe: The Social, Demographic and Financial Consequences of Europe's Ageing Population*, edited by R. de Bok, 18-26. Breda: PlantijnCasparie.

Köhler, Wim (2004). "'Ouderdom is groeiende chaos'. Interview met Rudi Westendorp." *M: het maandblad van NRC Handelsblad*, augustus, 34-39.

Korten, David C. (2000). *The Post-Corporate World: Life after Capitalism*. West Hartford, Connecticut / San Francisco, California: Kumarian Press / Berrett-Koehler Publishers.

Kuningas, Maris et al (2008). "Genes Encoding Longevity: From Model Organisms to Humans." *Aging Cell* 7: 270-80.

Kurzweil, Ray & Terry Grossmann (2004). *Fantastic Voyage: Live Long Enough to Live Forever.* Emmaus, PA: Rodale.

Legato, Marianne J. (2005). "Where Did I Leave My Keys? Men, Women, and Aging." In *Why Men Never Remember and Women Never Forget*, 199-221. Emmaus, PA: Rodale.

Lerner, Michael (2006). *The Left Hand of God: Taking Back Our Country from the Religious Right.* New York: HarperCollins, HarperSanFrancisco.

Mackenbach, Johan P. & Martijntje J. Bakker (2003). "Tackling Socioeconomic Inequalities in Health: Analysis of European Experiences." *The Lancet* 362, October 25: 1409-14.

Marmot, Michael (2005). "Social Determinants of Longevity and Mortality. SAGE Crossroads, 28 June [cited August 24 2006]. Available from http://www.SageCrossroads.net

——— (2004). *The Status Syndrome: How Social Standing Affects Our Health and Longevity.* New York, NY: Henry Holt, Times Books.

Marres, Noortje Suzanne (2005). *No Issue, No Public: Democratic Deficits after the Displacement of Politic*s. PhD Thesis. Universiteit van Amsterdam. Available from http://www.dare.uva.nl/document/17601

Mauron, Alex (2005). "The Choosy Reaper: From the Myth of Eternal Youth to the Reality of Unequal Death." *EMBO Reports* 6, Special Issue (July): 67-71.

McConnell, Charles & Leigh Turner (2005). "Medicine, Ageing, and Human Longevity: The Economics and Ethics of Anti-Ageing Interventions." *EMBO Reports* 6, Special Issue (July): 59-62.

McKibben, Bill (2003). *Enough: Genetic Engineering and the End of Human Nature.* London: Bloomsbury.

Melton, Lisa (2006). "A Family Affair: Genes and Ageing." In *Wellcome Focus 2006. Ageing: Can We Stop the Clock?*, edited by Giles Newton & Tom Freeman, 22-25. London: The Wellcome Trust.

Miller, Richard A. (2004). "Extending Life: Scientific Prospects and Political Obstacles." In *The Fountain of Youth: Cultural, Scientific, and Ethical Perspectives on a Biomedical Goal*, edited by Stephen G. Post & Robert H. Binstock, 228-48. Oxford: Oxford University Press.

Minicuci, N. et al (2004). "Disability-Free Life-Expectancy: A Cross-National Comparison of Six Longitudinal Studies on Ageing. The CLESA Project". *European Journal of Ageing* 1: 37-44.

Moody, Harry R. (1995). "The Meaning of Old Age: Scenarios for the Future." In *A World Growing Old: The Coming Health Care Challenges*, edited by Daniel Callahan et al, 9-19. Washington D. C.: Georgetown University Press.

Moody, Harry R., Arthur Caplan & Morton Kondracke (2004). Is Aging a Disease? SAGE Crossroads, January 22 [cited November 15 2004]. Available from http://www.SageCrossroads.net

More, Max (2004). "Superlongevity without Overpopulation." In *The Scientific Conquest of Death: Essays on Infinite Lifespans*, edited by Immortality Institute and Sebastian Sethe, 169-85. Buenos Aires: LibrosEnRed.

Neugarten, Bernice L. (1996). "Social Implications of Life Extension [1978]." In *The Meanings of Age: Selected Papers of Bernice L. Neugarten*, edited by Dail A. Neugarten 339-45. Chicago: The University of Chicago Press.

Nussbaum, Martha C. (2011). *Creating Capabilities: The Human Development Approach.* Cambridge, Mass.: The Belknap Press of Harvard University Press.

Oeppen, Jim & James W. Vaupel (2002). "Broken Limits to Life Expectancy." *Science* 296, no. 10 May: 1029 and 1031.

Olshansky, S. Jay et al (2006). "In Pursuit of the Longevity Dividend: What Should We Be Doing to Prepare for the Unprecedented Aging of Humanity?" *The Scientist*, 28-36.

Overall, Christine (2003). *Aging, Death, and Human Longevity: A Philosophical Inquiry.* Berkeley: University of California Press.

——— (2004). "Longevity, Identity, and Moral Character: A Feminist Approach." In *The Fountain of Youth: Cultural, Scientific, and Ethical Perspectives on a Biomedical Goal*, edited by Stephen G. Post & Robert H. Binstock, 286-303. Oxford: Oxford University Press.

Post, Stephen G. (2004). "Establishing an Appropriate Ethical Framework: The Moral Conversation around the Goal of Prolongevity." *Journal of Gerontology: Biological Sciences* 59A, no. 6 (June): 534-39.

Rawls, John (1999). *A Theory of Justice: Revised Edition.* Cambridge, Massachusetts: Belknap Press of Harvard University Press, 1999.

Scully, Jackie Leach (2004). "What Is a Disease? Disease, Disability and Their Definitions." *EMBO Reports* 5, no. 7: 650-53.

Smaling, Adri & Hans Alma (2010). "Zingeving en levensbeschouwing: een conceptuele en thematische verkenning." In *Waarvoor je leeft: studies naar humanistische bronnen van zin*, edited by Hans Alma & Adri Smaling. Amsterdam: SWP/Humanistics University Press: 17-39.

Smith, Jacqui et al (1999). "Sources of Well-Being in Very Old Age." In *The Berlin Aging Study: Aging from 70 to 100*, edited by Paul B. Baltes & Karl Ulrich Mayer, 450-471. Cambridge, UK: Cambridge University Press.

Stock, Gregory (2002). *Redesigning Humans: Our Inevitable Genetic Future.* Boston: Houghton Mifflin.

Stock, Gregory & Daniel Callahan (2004). "Point-Counterpoint: Would Doubling the Human Life Span Be a Net Positive or Negative for Us Either as Individuals or as a Society?" Journal of Gerontology: Biological Sciences 59, no. 6 (June): 554-59.

Stock, Gregory, Francis Fukuyama & Tom Miller (2002). The Genetic Future Is Now: Redesigning Humans vs. Regulating Science. Cato Institute, 15 March [cited 6 May, 2003]. Available from http://www.cato.org/events/transcripts/020315et.pdf

Stock, Gregory, Bill McKibben & Morton Kondracke (2003). Do We Want Science to Reinvent Human Aging? SAGE Crossroads, March 27 [cited 2004]. Available from http://www.SageCrossroads.net

van der Lucht, F. & J.J. Polder (2010). *Van gezond naar beter. Kernrapport van de Volksgezondheid Toekomst Verkenning 2010.* Bilthoven: Rijksinstituut voor Volksgezondheid en Milieu.

van Tongeren, Paul (1995). "Life Extension and the Meaning of Life." In *A World Growing Old: The Coming Health Care Challenges*, edited by Daniel Callahan et al, 28-38. Washington D. C.: Georgetown University Press.

Vincent, John A. (2006). "Ageing Contested: Anti-Ageing Science and the Cultural Construction of Old Age." Sociology 40, no. 4 (August): 681-98.

Warner, Huber et al (2005). "Science Fact and the Sens Agenda: What Can We Reasonably Expect from Ageing Research?" *EMBO Reports* 6, no. 11 (November): 1006-08.

Westendorp, Rudi G. J. & Thomas B. L. Kirkwood (2007). "The Biology of Ageing." In *Ageing in Society: European Perspectives on Gerontology*, edited by John Bond et al, 15-37. London: Sage.

Wijnbergen, S. J. G. van (2002). "Economic Aspects of an Ageing Population." In *Ageing in Europe: The Social, Demographic and Financial Consequences of Europe's Ageing Population*, edited by R. de Bok, 28- 39. Breda: PlantijnCasparie.

Wörner, Markus H. (2000). "Untimely Meditations on Ageing, the Good Life and a Culture of Friendship." In *Health Promotion: New Discipline or Multi-Discipline?*, edited by Ricca Edmondson & Cecily Kelleher, 205-18. Dublin: Irish Academic Press.

Part 3

Corporations, Governments and Democracy

10

The Influence of Government and Industry on Technological Research in the Netherlands: A Personal Experience

Alle Bruggink

Introduction

The traditional approach of curiosity driven fundamental research solely governed by scientific debate, followed by serendipitous or fully commercially inspired applications has already for some decades become history in the Netherlands. In particular for the technological sciences, experts from academia, government and industry interact and collaborate in varying models and intensity in formulating national research programs. Some of these models, such as the technological roadmap protocol, will be described in more detail, in particular regarding the role and influence of the parties involved. Mainly experiences in chemistry and life sciences research programs will be presented. Some of the tensions and dilemmas in the procedures will be discussed. Thus, the condition of the research to be precompetitive versus the desire of successful innovations and new start-up companies will be shown to be one of the reasons of the so called innovation gap. Also the predictable differences in opinion between curiosity driven researchers and commercially governed developers will be mentioned, emphasizing the absence of bad feelings between academia and industry. The minor role of societal organizations, including NGOs, in the early phases of research will be discussed, together with the much more strenuous discussions when applications and new enterprises are coming about.

Organizing a national research program in technical sciences: a successful start

The Netherlands have, in contrast to the European Union, no detailed protocols for setting up a national research program. At a cursory glance a national program is the result of a number of meetings and discussions between experts, followed by a program proposal which is commented and judged by some expert committee. International refereeing and reviewing is not a rule and the final judgment by an expert committee (followed by approval by the subsidizing authority) is often felt as unfair in case of criticism or an advice for disapproval because all experts have been consulted or involved in drawing up the program. In particular in a small country as the Netherlands it is easy to involve all colleagues in a national proposal, thereby making it difficult to find suitable and respected personal authorities for an expert committee. Through the years, however, the procedures have gained in quality, not the least by the growing number of experts becoming available through (early) retirements schemes, through the increasing involvement of industrial experts and the more frequent use of international expertise. In particular the role and participation of industrial experts have been fully accepted over a period of the last 25-30 years in which a gradual growth in mutual respect and openness with academic partners can be seen. Interestingly, not a single example of undesired commercial push in a national program can be pinpointed. In individual projects involving a limited number of parties and researchers there have been conflicts of interest that needed attention or arbitrage by program committees. This is a gratifying result for both the Dutch industry and academia and has contributed to our international strength and reputation in technical sciences. To many non-technical observers this successful collaboration between industry and university is a surprise. In particular the number of large international corporations (i.e. Shell, AKZO-NOBEL, DSM, Unilever, Philips) was of concern with respect to undesired commercial push in university research. Also the increasing political desire, and to a lesser extent also societal pressure, to exploit results of academic research and development in innovative applications in a knowledge based economy could easily lead to university research groups being guided by industrial experts and commercial goals. Nevertheless, the successful industry-university collaboration already holds for 25 years; up to a point where even industrial experts are chairing important institutions as the chemical department of the Netherlands Organisation for Scientific Research (NWO). Depending on definitions between 30 and 60% of research in technological sciences at Dutch universities is influenced by the national industry.

The key success factors

Key factors for successful collaboration, i.e. between industry and university, can be defined from different perspectives. From a general point of view they can be formulated as:

- Mutual respect and openness
- The presence of an agreed innovative need or benefit.

How did these KSFs apply to the national research programs in the technical sciences and what other, more intrinsic, factors made the success? Were there any drawbacks in the overall result and how could these be addressed? What was the role and influence of the government and societal parties as NGOs?

Parties, industry and university from the start of their organized collaborations in 1975-1980 were aware that mutual respect and openness about each other's motives and goals would be a key to success. A meaningful interpretation of this key, however, took a decade or so. It was in fact the condition set by the European Union that government subsidized research collaboration between universities and industry should be pre-competitive, that made the key to work. Thus the goals of academic research should be defined and agreed in terms of the advancement of science and the Dutch reputation in a given expertise. The goals of industrial R&D should be defined as a package of know-how needed to enhance the development of commercial applications. The applications as such should not be the goal; they are the sole responsibility of the industry, they can disappear or change over time and are by definition out of phase with the planning of academic research. Openness about each other's research results was in particular for the industry a difficult step. For academic researchers it is a way of life and a source for inspiration and new ideas. Industrial researchers had to find the balance between patents, trade secrets and publicity. Three factors made the balance: *i* the prerequisite of pre-competitive collaboration brought the commercial application often on a "safe" distance, *ii* all national research programs allow the industrial partners sufficient time, 6-18 months, to file for a patent on academic results before these are made public in a scientific paper or presentation (keeping patents applications secret until the granting decision is no longer possible) and *iii* the Dutch chemical and life sciences industries are hardly involved in hard mutual competition allowing an open door to in-house know-how for collaborative national partners (for collaborations in European programs this is often not the case). A very helpful factor in building mutual trust and respect between academia and industry is the large number of part-time professors from industry teaching and coaching research at a university. Up to 30% of professors in technical sciences at some universities are from industry, whereas some Dutch industries employ over 25 part-time professors in their ranks. This typically Dutch phenomenon enjoys great international admiration and is very instrumental to successful collaborative research.

The condition about the presence of an agreed need or benefit goes hand in hand with the need on mutual respect and openness. However, the way subsidizing organizations as national or regional governments or institutions are interpreting this success factor often poses a problem. Although not written in the program or the financing conditions they define success as a tangible application with an apparent economic benefit or an actually new entrepreneur. Although understandable from a short-term political or societal point of view, this desire is not in line with the formal conditions on pre-competitiveness and can hardly be met by the timings and dynamics of academic research and lead-times to successful commercial development. The acceptable political or societal time span between invention and innovation is no more than 4 to 6 years, whereas in most national programs this time is already needed to come to a mature invention. Industrial developments are often no further than patent applications by the time the national research program is finished. Tangible commercial or societal results are therefore no longer in the domain of the program organization and are, when success comes about, often greatly intermingled with other industrial (or academic) achievements. It is interesting to see that it is more due to government pressure than to industry when the push towards societal or commercial benefit goes beyond the agreed conditions. An additional argument for industry not to push too hard for commercial results is the limited amount of their money (25-50% of the total budget for national programs is coming from industry) involved in these programs and the much larger financial resources needed to bring an academic result to commercial success. All partners would be greatly disappointed when a scientific success could not be translated to commercial or societal benefit because industry would not have the resources to do so. In fact, the limiting factors from the industrial side are often much more the management time needed for in-depth scientific collaboration and the time and money that need to be allocated to further development when all collaborative projects came to success than the relatively limited amount of money going directly in the national research program.

The latter point is of particular importance for the role and participation of the SMEs (small and medium sized enterprises) in national research programs. So far their partnership is cumbersome, despite the great desire, by all parties including the SMEs themselves, for their presence. Contrary to the large corporations for all SMEs the required investment in money is limiting their participation. An attractive solution to this problem is capitalizing the time of involvement in the program of the SME's expert, preferably in a shared position between university and enterprise, and counting this as their financial contribution. A number of programs organized by the chemistry department of the Netherlands Organisation for Scientific Research in cooperation with the ministry of Economic Affairs together with a consortium of Dutch academia and industries are operating this formula. SMEs can fulfill their financial obligations up to 100% in men-hours, whereas larger industries have to contribute at least for 50% in cash.

Summarizing the Dutch model (see also table 1): industry and academia have a strong collaborative relationship based on mutual understanding and respect and aiming at pre-competitive research results with attractive options for innovation and economic benefit.

	Technical Experts		*Non-technical Experts*			*NGOs*	*Public*
	Academia	Industry	Econ.	Society	Environm.		
Preparations	xxx	xxx	x	x	x	x	-
Drawing proposals	xxx	xxx	-	-	-	-	-
Execution of the progam	xxx	x	-	-	-	-	-
Evaluations of the program	xxx	xxx	(x)	(x)	(x)	(x)	-
Evaluation of individual results	xxx	xxx	x	x	x	x!	-

Table 1: Involvement of stakeholders in national research programs

The government role is mainly restricted to financing and guarding the EU rules with an increasing desire for shorter-term results at societal or economic level. This looming paradox between pre-competitive collaboration and shorter-term benefit is at the basis of our innovation gap: the longstanding struggle to a quick translation of fundamental research results into innovative economic advantage.

Translation into a sustainable model: Technology Road Mapping

The key to technology road mapping is the development of a convincing view of the long term future about the theme of the proposed research program (forecasting) followed by a detailed summary and analysis of the actions to be taken in the short term aimed at reaching the goals and benefits of the long term vision (backcasting). The model is derived from business strategy dialogues in which scenarios for future market needs, societal developments and required or expected technologies play an important role. Ideally the road map process is done with four parties, industry, academia, government and societal representatives such as NGOs, guided by an experienced consultant group. Each party should be represented by at least 4 to 10 persons with a good mix of young potentials and proven experts. The primary role of each party is agreed upon beforehand, although all are equally participating in discussions and decision-making. Industry leads the what question and has to present a convincing long term, i.e. 25 years, business case for the subject; for example genomics, catalysis or separation technology. They should present a bright business potential for the topic together with its societal relevance and acceptance and a survey of the technological means needed. The business case should preferably be based on a SWOT[1] analysis and other suitable strategy defining techniques. The relevance for the Dutch economy and society should be shown together with our competitive advantages. There is a considerable risk in these very important first steps: the convincing view of the future should not be the result of the opinion of the "best man in class" (the most recognized professor or the company CEO) nor should it be the outcome of all the analytical tools applied to describe the case. The latter is too much an extrapolation of the present situation, while the former can be based on a personal view or hobby. The forecast should be shared by all parties and challenged throughout the entire process. Several months should be taken to reach the common picture of the future.

The academic parties are leading the how question, firstly by presenting a broad view about the future of their discipline followed by a survey of their ideas in delivering the required or desired technology fitting in the business case. They should also show how their views and ideas fit in their international reputations and how the national strongholds could become even stronger. Missing technologies and expertise have to be handled with great care. Most often they develop into a showstopper, unless unique ideas are available or international collaboration could lead to a win-win situation. A simple decision to remove a weak spot with some extra research should be avoided, as increasing international competition will hardly give us a chance. It is better to negotiate such topics for international (including European) cooperation with our strengths in our hands.

Government and NGOs are dealing with the why question whereby the government is looking for both national economic and societal benefit and NGOs zoom in on societal and environmental issues in national as well as international perspective. So far the role of these two parties has not been fully developed.

---

[1] SWOT = Strengths, Weaknesses, Opportunities and Threats

With the results of these three approaches in hand a first draft of the research program can be made in terms of which knowledge and results (of fundamental, developmental or applied character) will be needed at which time and in what detail and quantity. In that way a roadmap going from fundamental research all the way to business application can be set up for important goals or various parts of the program. A very important step at this stage is the quantification of the program goals. Thanks to the worldwide discussions on sustainability development improved tools have been developed in setting up quantified goals. Several technology assessment methods, including fully quantified life cycle analyses are nowadays available to describe the present and desired future state of science and technology. Improvements of at least a factor 10 in high level technology parameters are increasingly accepted in national research programs as a result of the aforementioned discussions on sustainability development. The simplified reasoning is that in order to compensate for an increasing world population and a much better participation of that population in higher levels of welfare and wellbeing, we have to utilize our resources at least a factor 10 more efficient in order not to damage (further) the state of our planet. If we have the desire to restore some environmental and societal damages from the past our scientists should surprise us with proven and applicable inventions allowing more than a factor 10 improvement in sustainability related parameters such as *i)* consumption of raw materials, *ii)* energy requirements, *iii)* emissions to the environment, *iv)* toxicity, safety and other risk potentials, *v)* area use and *vi)* economic and labor related issues. In particular the Dutch chemical parties have used this quantified approach to set up a number of successful national programs i.e. on catalysis, separation technologies, biosynthesis and development of new materials. Also in the energy sector technology road mapping is becoming a popular method to define long-term goals and to direct R&D programs, i.e. programs on hydrogen as energy carrier, solar energy developments and the utilization of biomass as energy source.

Towards a *lingua democratica*?

To what extent does the model presented above deviate from the classic model in which experts are setting the scene and define their own research programs? Not as much as would be anticipated. Indeed, all relevant questions, besides scientific and technological issues, are discussed and have become mandatory items in most national and EU R&D proposals. Chapters on economic, societal and environmental issues are common requirements. Quite often all questions are combined in a sustainability section and presented in terms of effects on People (societal issues), Planet (environmental issues) and Profit (economic issues). However, in most examples of national research programs, including the Netherlands Genomics Initiative, the experts in science and technology are drawing up all answers and are doing the final readings and editing. At best experts in societal, environmental and/or economic issues have been involved in the preparative discussions. Unlike industrial experts, technological scientists at universities are still very hesitant in consulting non-technological experts about their research ideas and proposals. In industry scientists are used to consult commercial, financial, safety, environmental and societal experts (mostly in that order but often in teams) in the milestone decisions of their programs and individual projects. The lack of experience in interdisciplinary teamwork at universities is at the basis of this situation. Personal achievements are still very much fundamental to our academic education and research systems (for numerous good reasons), preventing scientifically sound opinions on non-technological problems or potentials of technology directed research proposals and results. Very often in the introduction of PhD theses in physics, chemistry or biology benefits of the proposed research topic for the environment, for sustainability or some other relevant societal item, is presented, without scientific proof and mostly without a sound comparison between the existing situation and the proposals. After the research is completed and the results are known there is most times no more mention at all whether such beneficial effects and the quality thereof have been achieved. There is no good reason not to involve non-technological experts, such as in economy, sociology, law and ethics, or whatever relevant discipline, in starting or finishing a PhD thesis. The same holds for setting up national research programs for which large amounts of public money are needed and whose results can have huge societal impact. The technological experts should still be in the leading position, but they could improve a lot on the quality of the non-technological aspects of their proposals by having the appropriate chapters conceived by the best colleagues in the appropriate discipline.

For the involvement of NGOs the picture is slightly different and as experience has shown also somewhat more complex. Practice has shown that NGOs can participate very well in preparative discussions of research proposals. Mostly they agree on the research goals, simply because the goals are always directed at improving an existing situation and formulated in terms "nobody could disagree with". The problems arise when results are achieved and (commercial) exploitation or potential is at the table. At that point, NGOs can start their real role by disputing the benefits of the results and the (unwanted or unexpected) side effects. Mostly, the technology-oriented scientist is rather disappointed about the opinion of the NGO at this moment of his project. After the consent of the NGO with the research program and its goals, the researcher anticipated applause at the time of the results. The researcher took time and effort to reach his results, while the NGO was just sitting still and only asking critical questions afterwards. The researcher almost feels betrayed by the NGO, forgetting they can only play their role with concrete results at hand. It is interesting to see how such criticizing comments are accepted when stemming from economic reasons (i.e. the product is too expensive or there is no customer). When the research results are stemming from teamwork, as mostly in industry, the criticizing role of all participants, including from experts in safety, environmental or societal topics, is accepted. But again, critical comments of NGO representatives, having not been involved in the risks and efforts of the research, are received with mixed feelings. A more satisfying position of NGOs could be reached by *i)* continuing communication between researchers and the NGO, *ii)* acceptance by the researchers that NGOs can only play their role in the development chain when applicable results are available and *iii)* an equal level of scientific quality in presenting the research results and the critical comments of the NGO.

In conclusion (see also Table 2), a *lingua democratica* is still rather remote in setting up national research programs, but meaningful developments are underway. In industry a further development of interdisciplinary teamwork, including team members from outside the own industry, could be a positive next step. In academia the full involvement of the best experts in the appropriate non-technological disciplines, allowing an equal level of scientific soundness in technological and non-technological issues, would be a welcome improvement. Such an *a priori* involvement, followed by continuing discussions during the research progress, could very well be a prelude to successful interdisciplinary teamwork as already practiced in industry. The somewhat emotional bias against NGOs should be taken away by a mutual acceptance of the individual roles in the development process and by an equal level of scientific quality in presenting research results and critical comments, both for technological and non-technological matters.

	Technical Experts		*Non-technical Experts*				*NGO's*	*Public*
	Academia	Industry	Econ.	Society	Environm.			
Preparations	xxx	xxx	x	x	x		x	-
Drawing proposals	xxx	xxx	*x*	*x*	*x*	*Step 1*	-	-
Execution of the progam	xxx	x	**x**	**x**	**x**	**Step 2**	**(x)**	-
Evaluations of the program	xxx	xxx	*x*	*x*	*x*	*Step 1*	*x*	-
Evaluation of individual results	xxx	xxx	x	x	x		*Step 3* **x**	**x**

Table 2: Involvement of stake holders in national research programs; recommendations for improvements towards a "lingua democratica"

Further reading

1) http://www.nwo.nl/gw for the following national research programs:
- De Maatschappelijke Component van het Genomics-onderzoek
- Ethiek, Onderzoek & Bestuur
- Maatschappelijk Verantwoord Innoveren (MVI)

2) http://www.genomics.nl/GenomicsInSociety.aspx

3) http://www.genomics.nl/Valorisation.aspx

4) http://www.nwo.nl/nwohome.nsf/pages/NWOA_6P69LY for a number of national technology programs related to chemistry, technology and sustainability.

5) www.nwo.nl/actsmeansbusiness for discussions on technology and society.

6) http://www.catchbio.com/ as an example of a technology program in relation to sustainability development.

7) *Catalysis, key to sustainability*, Technology Roadmap Catalysis, Ministry of Economic Affairs, October 2001. See also www.ez.nl and www.technologyroadmapping.com

8) http://dsti.nl/roadmap as an example of technology roadmapping.

9) *Publieksonderzoek Genomics*, Stichting Consument en Biotechnologie, Den Haag November 2002; see also "Monitor voor het Regie-Orgaan Genomics" at http://www.genomics.nl/GenomicsInSociety.aspx

10) *Silent Revolution, Dutch industry and Dutch government working together for a better environment*, Ministry of VROM, The Hague, The Netherlands, October 1998.

11) *Translating Sustainability Strategy into Action, The Chemical Industry*, B. Beloff, M. Line and D. Tanzil (Eds.), Wiley-Interscience, 2005, ISBN 13: 978-0-471-6445-3

12) *Renewable-Based Technology, Sustainability Assessment*, J. Dewulf and H. van Langenhove (Eds.), Wiley-Interscience, 2006, ISBN 13 978-0-470-02241-2

11

Consumer Genomics Companies and Their Appeal to Empowerment

Anders Nordgren

Introduction

During the last few years, we have seen an avalanche of companies offering DNA testing online "direct-to-consumer". Some of these consumer genomics companies offer non-health-related DNA information such as genetic ancestry tracing and ethnicity analysis (e.g. DNA Worldwide). Others provide health-related DNA information, e.g. disease risk testing, pharmacogenetic testing and nutrigenetic testing (e.g. DNA Direct). A third category provides both health-related and non-health-related information. Some companies in this category – for example, deCODEme, 23andMe, Knome, and Navigenics– offer whole-genome scans, not merely information about a limited number of DNA sequences.

In the web-advertisement of some companies, two interesting appeals are made, explicitly or implicitly. The first is an appeal to identity. The very names of the "Me-companies" - 23andMe, deCODEme and Knome - indicate this. The DNA information provided by the companies is suggested to be pertinent to personal identity. Together with Eric Juengst, I have elsewhere analyzed and discussed this appeal.[1] In the present chapter, I will discuss another appeal, namely the appeal to personal empowerment. DNA information direct-to-consumer is suggested to empower the consumer. I will analyze what this appeal could mean more precisely and discuss whether it is justified.

[1] Nordgren & Juengst, 2009

The appeal to empowerment

Let me start with a few examples of this appeal to empowerment. Navigenics maintains (2011):

> Navigenics is the leading provider of clinically guided genetic analysis. Our goal is to empower you with genetic insights to help motivate you to improve your health.

On the web-site of 23andMe (2011), one can read:

> 23andMe was founded to empower individuals and develop new ways of accelerating research.

In a similar vein, deCODEme states on its website (2011):

> Getting to know your personal genome will empower you and provide you with a road map to improve your health.

In these quotations, we see an explicit appeal to empowerment. There are also other statements on these companies' websites that can be interpreted as making an implicit appeal to empowerment, i.e., that do not use the terms "empowerment" or "empower" but talk about similar ideas.

This appeal to empowerment has received some academic attention,[2] but a thorough analysis is lacking. The appeal raises three serious issues. The first is conceptual. What does the appeal to empowerment mean? The second issue concerns justification. Is the appeal to empowerment justified? Finally, there is an empirical issue. Do the consumer genomics companies in fact empower their consumers and are there differences in this regard in different parts of the world – for example, in the US compared to Europe – due to differences in healthcare systems or other societal differences? The empirical issue is beyond the scope of this chapter; I will focus on the conceptual issue and the issue of justification.

What is empowerment?

The concept of empowerment has been discussed in three theoretical contexts. The first is critical social theory and emancipatory theory

[2] Bowen et al. 2005

(e.g., feminism). Empowerment is understood as improvement of the living conditions of minorities, women, patients and other oppressed groups. The second type is organization theory, in which empowerment is the delegation of power and giving people opportunity to take action. Third, there is social psychological theory. In this context, empowerment concerns the individual's development.[3]

Tengland has provided a conceptual analysis of empowerment focusing on the concept's use in relation to health promotion.[4] I find his analysis very useful in trying to understand the appeal to empowerment made by the consumer genomics companies.

Tengland makes a distinction between empowerment as a goal and empowerment as a process or approach. Empowerment as a goal can be understood in terms of control (for example, gaining mastery of one's own health or one's own life) or internal resources (for example, obtaining knowledge, becoming more autonomous, changing one's self-image). Empowerment as a process or approach involves the participants in problem-formulation, decision-making and action in a way that makes the experts having to withdraw some of their control and power.[5]

What does the companies' appeal to empowerment mean and how is it justified?

It appears that empowerment is a very elusive concept that can be understood in many different ways. One might suspect that this elusiveness in combination with its positive connotation makes it very useful in the rhetoric of the consumer genomics companies. The question is what the concept could mean more precisely in this context. I will investigate possible meanings of the companies' appeal to empowerment by using Tengland's distinction between empowerment as a goal (in the sense of control or internal resources) and empowerment as a process or approach.

---

[3] See Kuokkannen & Leino-Kilpi, 2000, for a more elaborate presentation and useful references

[4] Tengland, 2007

[5] Tengland, 2007

As a first approximation, the appeal (A) to empowerment can be articulated as follows:

> A: DNA information direct-to-consumer empowers the consumer.

With "DNA information" I mean DNA information that is based on samples from the consumer and that is at least to some extent explained to the consumer. With "direct-to-consumer" I mean that the DNA information is ordered directly by the consumer from the company without any intermediary such as a family doctor or other physician.

However, a closer look immediately suggests that there are two possible versions of this appeal to empowerment, one health-related appeal (HA) and one non-health-related (NHA):

> HA: Health-related DNA information direct-to-consumer empowers the consumer.
>
> NHA: Non-health-related DNA information direct-to-consumer empowers the consumer.

Empowerment and health-related DNA information

Let me start with possible interpretations of the appeal to empowerment as a goal to be reached by health-related DNA information direct-to-consumer. If we apply Tengland's analysis (with some modification), we get the following two main interpretations or aspects:

> HA1: Health-related DNA information direct-to-consumer increases the consumer's control in terms of gaining mastery of his/her own health or life.
>
> HA2: Health-related DNA information direct-to-consumer increases the consumer's internal resources in terms of obtaining knowledge, becoming more autonomous, or changing his/her self-image.

Examples of both interpretations (or aspects) can be found on the websites of consumer genomics companies. Control (HA1) is mentioned explicitly by 23andMe: "Take charge of your health and wellness today" (2011). And knowledge (HA2) is mentioned explicitly in the quotation from Navigenics above: 'Our goal is to empower you with genetic insights to help motivate you to improve your health' (2011).

How can health-related DNA information direct-to-consumer empower consumers, whether understood in terms of increased control or internal resources? What is the proposed justification for the appeal? The major arguments are found here and there on the websites. Empowerment in each of the above senses may be accomplished by obtaining DNA information about disease risks (clinical genetics), because this may lead consumers to start taking certain drugs, change lifestyle or avoid having children (in order not to genetically transmit the disease risk to these children). It may be achieved by receiving DNA information about drug response (pharmacogenetics), because this may lead the consumers to take certain drugs but avoid others. And it may be realized by obtaining DNA information about nutritional needs (nutrigenetics), because this may lead consumers to make certain changes concerning diet and lifestyle.

There is also a third interpretation of the term "empowerment", namely as a process or approach to be realized by health-related DNA information direct-to-consumer:

> HA3: Health-related DNA information direct-to-consumer increases the involvement of the consumers in taking responsibility for their own health and decreases to some extent the involvement of genetics experts, family doctors, and other physicians in the regular healthcare system.

For example, 23andMe provides this reference to increased consumer involvement: 'Make better health decisions by learning your genetic risks' (2011). Moreover, the company states that "23andMe provides you with all your data and will never withhold it from you" (2011) (meaning that conventional genetics experts might want to (my comment)).

How can health-related DNA information direct-to-consumer empower consumers in the sense of increasing consumer involvement? What reasons are given? The main argument can be seen on the web-site of DNA Direct (in a press release):

> With just over 2,000 genetic counselors nationwide and even fewer medical geneticists, most people don't have acces to genetic testing. DNA Direct is bridging this gap

> to provide people with a resource to better understand, evaluate and, if they choose, work with their physicians to better manage their health and healthcare decisions.

In short, the reason is that there are very few medical geneticists and genetic counsellors available, which has the consequence that most people do not have access to genetic testing and its benefits. Health-related DNA information direct-to-consumer meets this need. It should be noted, however, that DNA Direct states that it does not want to compete with the consumers' physicians but rather work together with them.

Empowerment and non-health-related DNA information

Empowerment is nowhere on the websites of the consumer genomics companies explicitly mentioned in relation to non-health-related DNA information. However, two of the companies quoted above as referring to empowerment – 23andMe and deCODEme – provide not only health-related DNA information but also non-health-related. This means that it is at least possible to interpret them as claiming implicitly that also this kind of information may empower the consumers. Actually, I will try to show that it is possible to argue that non-health-related DNA information direct-to-consumer empowers consumers (although I will also argue further below that this is not unproblematic).

Applied to non-health-related DNA information, Tengland's conceptual scheme suggests the following interpretations (aspects):

> NHA1: Non-health-related DNA information direct-to-consumer increases the consumer's control in terms of gaining mastery of his/her own life.
>
> NHA2: Non-health-related DNA information direct-to-consumer increases the consumer's internal resources in terms of obtaining knowledge, becoming more autonomous, or changing his/her self-image.

Now, how can non-health-related DNA information direct-to-consumer empower consumers in these senses? No explicit reasons can be found on the websites, which is not to be expected since they do not explicitly, but only possibly implicitly, link empowerment to non-health-related DNA information. However, the main reasons might be reconstructed as follows. Non-health-related DNA information might empower a consumer by supporting the consumer's self-image and subjectively experienced social identity such as identity linked to paternity, extended family relationships, ancestry, ethnicity, or uniqueness. For example, the consumer might discover that he is in fact the father of the child, that the person A is actually his cousin, or that he is truly of Nigerian ancestry. Another reason could be that non-health-related DNA information might change the self-image in a way that is experienced as positive. For example, the consumer didn't previously believe that he was that father of the child but now this is confirmed and he is happy about it. Similarly, he might be happy finding out that person A is his cousin despite the fact he didn't believe so. Moreover, he might be happy discovering that he is actually of Nigerian ancestry although he didn't expect it. All these insights might have profound practical consequences and are not merely a matter of subjective experience. The insights might lead the consumer to take action is certain respects. Therefore it is truly a matter of empowerment – or at least so the argument can be reconstructed.

In a third interpretation, empowerment is seen as a process or approach to be realized by non-health-related DNA information direct-to-consumer:

> NHA3: Non-health-related DNA information direct-to-consumer increases the involvement of the consumers in taking responsibility for aspects of their own life related to this information.

It should be noted that in articulating this interpretation, I have skipped the reference in HA3 to decreased control and power by genetics experts, because earlier there was no possibility to get to know these things through these experts, either. Now the consumer has the opportunity to obtain this information directly from the consumer genomics companies. Actually, this fact constitutes the reason why this should be considered an empowerment. The consumer gets involved in something that he/she could never be involved in before.

Does the proposed justification of the appeal to empowerment hold water?

It seems that the appeal to empowerment might be a useful rhetoric tool, but is its proposed justification acceptable?

Empowerment and health-related DNA information

Does the health-related DNA information direct-to-consumer empower the consumers in the sense that it leads to increased control or internal resources? We have seen the *pro* arguments above referring to the benefits of obtaining DNA information about disease risk, drug response, and nutritional needs. However, there are also important *con* arguments.

First, the information may be much less informative than advertised. An example can be found in the investigation of four companies offering nutrigenetic testing carried out by the US Government Accounting Office (GAO). Fourteen "different" DNA samples were sent to the companies, although in fact they originated from only two individuals (a 48-year-old man and 9-month-old baby girl). It turned out that the recommendations from three of the companies simply mirrored the fictitious information that followed each sample. Those who were smokers were recommended to stop smoking etc. The companies maintained that the results would include no medical predictions, however all 14 results did include predictions that could be interpreted as diagnoses. The GAO report concluded that none of the results from the companies contained scientific support.[6] This report was published in 2006. In 2010, the GAO published a new report with similar critical conclusions.[7] Inadequate information, lack of validation of tests, or exaggerated or unsupported claims, may have serious consequences.[8] All this may influence the consumer to believe that he has a disease or is at high risk of developing a disease, although this is not in fact the case. The consumer may be misinformed, and this means that he or she is not empowered, quite the opposite. Certainly, many companies have substantial disclaimers. 23andMe (2011) states that:

---

[6] US Government Accounting Office, 2006
[7] US Government Accounting Office, 2010
[8] Hudson et al. 2007

> 23andMe Services are for research, informational, and educational use only. We do not provide medical advice.

Moreover, deCODEme (2011) stresses that:

> The Genetic Scan product is for informational purposes only, is not medical advice, and is not a substitute for professional medical advice, genetic coundeling, diagnosis, or treatment.

Disclaimers like these might seem responsible, but they indicate that the DNA information may not be as informative as one might think at first glance.

Second, there is a risk for misinterpretation of information. The consumer may not understand the limitations of the tests. The risk and severity of a disease may be exaggerated but also underestimated, and it might be difficult to distinguish widely used and accepted tests from invalid ones. For example, the consumer may get test results for a common disease in terms of relative risk compared with someone of the same age or gender in the general population. However, from some companies she might also get results in terms of absolute risk to develop the disease over lifetime. This means that she might have a low relative risk but a high absolute risk. This distinction between relative and absolute risk may not be properly understood. The customer may develop a false sense of security, worry unnecessarily, or neglect taking proper preventative action.[9] The consequence of misunderstanding will be that the customer is not empowered.

In conclusion, inadequate information and misinterpretation of information might decrease the consumer's control or internal resources, rather than increase them. The goal of empowerment will not be accomplished.

If we turn to empowerment as a process or approach, do we have reason to believe that the DNA information leads to increased involvement of the consumers and decreased control and power of genetics experts, family doctors, and other physicians in the regular healthcare system? In a sense the consumer involvement will certainly increase and the control of doctors decrease, but if the consumer is

---

[9] Kaye, 2008

misinformed or misinterprets the information, he or she will still not be empowered. In other words, if the goal of empowerment is not reached, empowerment as a process or approach will be of little value. Moreover, there is a risk that the genetic services of the consumer genomics companies empower the companies – in terms of getting control over their customers (and profiting on them) – at the expense of the regular healthcare system!

Empowerment and non-health-related DNA information

What about non-health related information direct-to-consumer, does this information empower the consumers in the sense that it leads to increased control or internal resources? The arguments for this being the case were presented above, namely that this information may support the consumer's self-image and subjectively experienced social identity, or that it may change the self-image in a way that is experienced as positive. However, there are also several problems.

Before we turn to these problems, however, let me point out that non-health-related DNA testing is not merely recreational, but may – as Bolnick et al stress – have serious consequences. It may reshape the customers' individual identities. They may experience emotional distress if test results are unexpected or undesired. In the US, the results may change how the customers report their race or ethnicity on governmental forms, college applications, job applications and medical questionnaires. Moreover, the tests have led African-Americans to visit and financially support specific African communities. And some consumers hope to obtain native-American tribal affiliation (and benefits like financial support, housing, education, healthcare and affirmation of identity) or to challenge tribal membership decisions.[10]

The first objection to the appeal to empowerment is that there is a risk for inadequate information due to scientific limitations. In genetic ancestry tracing, for example, two main methods are used in order to find particular mutations that have occurred at different points of time during human history. One method focuses on the mitochondrial DNA, the other on Y chromosome DNA. The mutations define certain populational "haplogroups". However,

---

[10] Bolnick et al. 2007

membership in a particular haplogroup doesn't say much about the consumer's genealogy in detail. Moreover, both methods provide information about only one line of descent (i.e., one ancestor per generation) out of all those that have contributed.[11] In addition, however, there is another method with autosomal markers that estimates the probability of biogeographical ancestry. But the problem with this method is that many potential patterns may be compatible with a particular result.[12] If the statistical limitations are not pointed out the information is inadequately presented. Moreover, it is clear that the statistical limitations can easily be misinterpreted by the customer.

The second *con* argument is that non-health-related DNA information may undermine social identity, not only support it. It may change the self-image in a negative way, leading to much emotional distress. You believed that you were the biological father, but you were not. You thought that you fulfilled the genetic requirements for belonging to a certain ethnic group, but you didn't etc.

Third, the companies do not clarify – or at least do not state clearly enough – that DNA information is not the only source for family identity and ethnic identity. An individual may belong to a family because he has been adopted or despite not being a genetic father, a social one. And an individual may be a member of an ethnic group even if she does not fulfil all the common genetic requirements. When social and genetic perspectives on identity conflict the question arises: what should carry the most weight: social attribution or genetic information?[13] In many cases, it is appropriate that the social attribution is considered the most important. If you state that you belong to a particular ethnic group you belong to that group, even if your deep ancestors (according to the DNA information) were not part of the parental population associated with that ethnic group.[14] In other contexts, however, the genetic information carries the most weight, the clearest example being judicial paternity trials.

---

[11] Shriver & Kittles, 2004
[12] Shriver & Kittles, 2004
[13] Nordgren 2008a; Nordgren & Juengst, 2009
[14] Bolnick et al. 2007; Frudakis, 2008

Fourth, even if the DNA information supports social identity in certain senses, this may actually only reinforce social stereotypes. The reason is that the social identity in question – for example, being 'black' or being 'white' – represents a stereotypic way of classifying people that we should get rid of. It is simply socially and politically irrelevant. We are all human beings and that is what should count morally, socially and politically.

This brings us to the fifth objection, namely that there are difficult conceptual problems concerning the relation of genetics, on the one hand, and race and ethnicity, on the other. This problem is actually illustrated on the companies' websites. For example, Genelex offers ethnicity testing. In contrast, Oxford Ancestors states that '[t]here is no genetic basis for ethnicity or race'.[15] Frudakis, Chief Scientific Officer at the company DNAPrint Genomics, can be interpreted as representing a middle course. He states that genetic ancestry testing reports only one aspect of 'race' and 'ethnicity'. Admixture is the rule rather than the exception. Therefore testing rather supports the idea that 'human-derived notions of "race" are based on the subjective and ever-changing concepts of social and political identity'.[16]

I conclude that there are many problems associated with non-health-related DNA information direct-to-consumer. It might be less informative than advertised because of scientific limitations and it may be open to misinterpretation. This means that the customer in the end might not be empowered by this information.

Finally, the issue of empowerment as a process or approach: do we have reason to believe that non-health-related DNA information direct-to-consumer leads to increased involvement of the consumers? In one sense 'yes', because no such information has ever before been available. However, as with health-related DNA information, if the customer is misinformed or misinterprets the information, he or she will still not be empowered. If the goal of empowerment is not accomplished, the value of empowerment as a process or approach will be limited or even non-existent.

---

[15] Oxford Ancestors, 2008

[16] Frudakis, 2008

The need for legal regulation

Whether or not this development of direct-to-consumer DNA testing leads to empowerment depends on whether the DNA information is adequate, based on validated tests, and properly explained, and on the measures taken to prevent misinterpretation. To ensure high quality of the genomic services legal regulation seems necessary.

A step in this direction can be seen in the *Additional Protocol on Genetic Testing*, issued by the Council of Ministers of the Council of Europe in 2008. It applies to all genetic risk testing, whether provided publicly or privately. It states three particularly important things. First, it requires that the clinical utility is established before the test is offered. Second, the test may only be performed under individualized supervision. Third, it requires all genetic testing to be followed by genetic counselling.[17]

Since the consumer genomics companies have been criticized both with regard to the premature nature of some of the tests and the lack of individualized supervision and mandatory genetic counselling, this regulation might have vast practical implications. A problem is that it is unclear to what extent the regulation is applicable to the kind of services offered by these companies. It is obviously applicable to the conventional type of genetic risk testing provided by, for example, DNA Direct. However, it might not be applicable to the type of health-related testing offered by, for example, 23andMe and deCODEme. As we saw in the disclaimers quoted above, 23andMe stresses that its services 'are for research, informational, and educational use only' and deCODEme points out that 'the Genetic Scan product is for informational purposes only, is not medical'. On the other hand, many customers probably interpret the tests as medical, and if they were not these people would perhaps not be interested in the first place.

The *Additional Protocol* represents the first regulation in Europe in this area. However, it will only affect countries that have signed the *European Convention on Human rights and Biomedicine*.[18] The UK, for instance, has not signed it. Moreover, even for those countries that have signed it, it is still optional and needs to be implemented into

[17] Council of Europe, 2008

[18] Council of Europe, 1997

national legal regulation for full effect. In the US, the situation is even worse. At present, there exists no appropriate regulatory system at the federal level. In the states of California, New York and some other states, however, some action has been taken against consumer genomics companies. The states require all tests to be requested by physicians.[19] deCODEme, for example, respects this regulation and 'may omit certain genetic risk information to residents of states where providing such information is restricted'.[20]

However, legal regulation is not sufficient. The responsibility for working out the details in practice lies with the geneticists. They need to assume responsibility for how to present their genomic services and for how to provide individual supervision and genetic counselling.

23andMe and deCODEme are rather good examples of how the presentation should be made, stressing the complex interaction of genes, environment and lifestyle. However, the limitations of the tests need to be emphasized even more strongly than is done on their websites.

Regarding individualized supervision and genetic counselling, it is not enough to recommend the customer to seek the advice of her physician or other qualified health provider (like 23andMe[21]) or even to offer it at no extra cost but leaving it to the customer to take the initiative (as deCodeme[22]). As stated by the *Additional Protocol*, individual supervision and genetic counselling should be mandatory when it comes to health-related DNA information. In line with the *Additional Protocol*, it is also important that the counselling is provided within a policy framework of 'non-directiveness'.[23] The clients, not the physicians or genetic counsellors, are to decide about whether or not to take a test and about what to do on the basis of test results. This is not a problem with regard to the consumer genomics companies, since this fits very well with the companies' general business ideas. However, as I have argued elsewhere, it is not enough that the information is given in a neutral manner (to the extent this is

---

[19] Kaye, 2008
[20] deCODEme, 2011
[21] 23andMe, 2011
[22] deCODEme, 2011
[23] Council of Europe, 2008

possible) and that the decisions are left to the clients. It is also vital that genetic counsellors give assistance in decision-making by challenging the views of the clients in order to help them discover their true value commitments and reach a well-considered decision. Moreover, when tests indicate a high disease risk, clients may experience a psychological crisis. Therefore it may sometimes also be necessary to provide psycho-social support.[24] Only in this way DNA information can truly empower the consumer. A practical problem is of course the shortage of genetic counsellors. Traditional face-to-face counselling can perhaps not be achieved. But in the Internet era, the use of web cameras is a possibility. Anyhow, much effort is needed in order to develop new ways of providing counselling services.

Democratization of the access to personal genomic information?

Finally, I will briefly discuss another issue raised by the above analysis of empowerment, namely whether DNA information direct-to-consumer democratizes access to DNA information. Foster and Sharp express the following expectation:

> Although the initial whole-genome online offerings are marketed primarily to affluent clients, the Internet has the longer-term promise of democratizing access to personal genomic information. For example, initial experiences with self-selecting clients could be used to pilot more effective and direct ways of delivering genetic education and counselling services to much larger numbers of people accessing their personal genomic information than are currently seen by genetic counsellors or clinical geneticists.[25]

In order to discuss this issue, we need to clarify what 'democratization' means. Foster and Sharp seem to mix quantitative and qualitative aspects. The quantitative aspect is that 'much larger numbers of people' get access to DNA information than in regular health care. The qualitative aspect is indicated by their talk about

---

[24] Nordgren, 2008b

[25] Foster & Sharp, 2008

'more effective and direct ways of delivering genetic education and counseling'. They seem to presume that merely providing DNA information is not sufficient; the information needs to be accompanied by appropriate education and counselling. Thus, Foster and Sharp seem to mean that access to personal genomic information will be democratized if and only if much larger numbers of individuals get access to this information and this access is combined with appropriate education and counselling.

Are Foster and Sharp right? Can the services of the consumer genomics companies be expected to lead to a democratization of the access to personal genomic information?

An argument in favour is that already at present more people have got access to DNA information than only a few years ago and that we have no reason to believe that this development will stop. On the contrary, it can be expected to continue for the foreseeable future. Moreover, the consumer genomics companies will probably be more or less forced by the consumers and the government to try to find new ways of handling the need for genetic education and counselling. Given this pressure, they will probably succeed in doing so.

A *con* argument is that the increased access to DNA information is not democratic in a true sense, because only those who can afford it get access. It is an access for the wealthy and rather wealthy only, not for each and everyone. Even when the vision of 'the 1,000 dollar genome' is realized, it will be too expensive for many people.

It is difficult to assess these arguments. I agree that the consumer genomics companies broaden the access to DNA information and in that sense further democratization. However, democratization in a true sense will probably be accomplished only when each and everyone gets free access to their personal DNA information within the regular health care system (health-related DNA information) and from public genomic service centres (non-health-related DNA information). However, the services of the consumer genomics companies might be an important intermediary step in this direction.

Conclusion

I have found that the term "empowerment" can be understood in different ways and that this ambiguity in combination with its positive connotation makes it very useful in the rhetoric of the consumer genomics companies. The appeal to empowerment can be criticized, however. DNA information direct-to-consumer – whether health-related or non-health-related – does not necessarily empower the consumer, because of the risk for misinformation and misinterpretation. Legal regulation is needed to counteract these risks. Genetic counselling should be made mandatory. Although DNA information direct-to-consumer may be viewed as a step towards democratization of the access of DNA information, the ultimate goal should be that each and everyone gets free access to their personal DNA information within the regular health care system (health-related DNA information) and from public genomic service centres (non-health-related DNA information).

Literature

23andMe, 2011. Available at: https://www.23andme.com (accessed February 9, 2011).

Bolnick DA et al. (2007). Genetics. The Science and Business of Genetic Ancestry Testing. *Science* 318 (5849): 399-400.

Bowen. DJ, Battuello, K. M., Raats, M. (2005). "Marketing Genetic Tests: Empowerment or Snake Oil?"*Health Education & Behavior* 32: 676-685.

Council of Europe (1997). Convention for the protection of Human Rights and the dignity of the human being with regard to the application of biology and medicine: Convention on Human Rights and Biomedicine. CETS No. 164. Available at: http://conventions.coe.int (accessed February 9, 2011).

Council of Europe, (2008). *Additional Protocol to the Convention of Human Rights and Medicine, concerning Genetic Testing for Health Purposes.* CETS No. 203. Available at: http://conventions.coe.int (accessed February 9, 2011).

deCODEme (2011). Available at: www.decodeme.com (accessed February 9, 2011).

DNA Direct (2011). Available at: www.dnadirect.com (accessed February 9, 2011).

DNA Worldwide (2011). Available at: www.dna–worldwide.com (accessed February 9, 2011)

Foster, M.W. & Sharp, R.R., (2008). "Out of sequence: How consumer genomics could displace clinical genetics". *Nature Reviews Genetics* 9 (6) : 419.

Frudakis, T. (2008). "The Legitimacy of Genetic Ancestry Tests". *Science* 319 (5866): 1039-1040.

Genelex (2011). Available at: www.healthanddna.com/ancestry-dna-testing/ethnicity-dna-testing.html (accessed February 9, 2011).

Hudson, K. et al. (2007). "ASHG Statement on Direct-to-Consumer Genetic Testing in the United States". *The American Journal of Human Genetics* 81: 635-637.

Kaye, J. (2008). "The regulation of direct-to-consumer genetic tests". *Human Molecular Genetics* 17 (R2): R180-R183.

Knome (2011). Available at: www.knome.com (accessed February 9, 2011).

Kuokkannen, L. & Leino-Kilpi, H. (2000). "Power and empowerment in nursing: three theoretical approaches". *Journal of Advanced Nursing* 31 (1): 235-241.

Navigenics (2011). Available at www.navigenics.com (accessed February 9, 2011).

Nordgren, A. (2008a). "Genetics and Identity". *Community Genetics* 11: 252-267.

Nordgren, A. (2008b). "Prenatal genetic counselling: Conceptual and ethical issues". In: Leuzinger-Bohleber M, Engels E-M, & Tsiantis J (Eds.). *The Janus Face of Prenatal Diagnostics: A European Study Bridging Ethics, Psychoanalysis, and Medicine.* London: Karnac: 307-325.

Nordgren, A. & Juengst, E.T. (2009). "Can Genomics Tell Me Who I Am? Essentialistic Rhetoric in Direct-to-Consumer DNA Testing". *New Genetics and Society* 28 (2): 157-172.

Oxford Ancestors (2011). Available at: www.oxfordancestors.com (accessed February 9, 2011).

Shriver, M.D. & Kittles, R.A., (2004). "Genetic ancestry and the search for personalized genetic histories". *Nature Reviews Genetics* 5: 611-618.

Tengland, P.A., (2007). "Enpowerment: A goal or a means for health promotion?" *Medicine, Health Care and Philosophy* 10: 197-207.

United States Government Accountability Office (2006). "Nutrigenetic Testing: Tests Purchased from Four Web Sites Mislead Consumers"". GAO-06-977T. Available at: www.gao.gov/new.items/d06977t.pdf (accessed February 9, 2011).

United States Government Accountability Office (2010). Direct-To-Consumer Genetic Tests: Misleading Test Results Are Further Complicated by Deceptive Marketing and Other Questionable Practices". GAO-10-847T. Available at: www.gao.gov/products/GAO-10-847T (accessed February 9, 2011).

12

Fundamental Uncertainty as an Issue for Deliberative Democracy

Michiel Korthals

Introduction

The knowledge of the human genome has great potential for human health. It may help us to understand the origins of diseases and help us to tackle diseases more precisely by using medicines that work at DNA level. Also for the nutritional sciences it possesses immense possibilities. For decades now epidemiologists have tried to find links between health and food. Research at genome level may help to reveal the relation between food and health. It could find the origins of diseases and help to prevent them. It may prove the healthy influence of nutrients. The DNA was thought to be the blueprint of our human body, concealing the secrets of life. If this blueprint would be uncovered we would 'know' for certain. 'We' would be able to solve the mysteries of our body.[1]

However, since the major part of the human genome has been sequenced in 2000, the positive idea is shattered into pieces. The potential of genomics is overshadowed by the enormous complexities surrounding the genome and the possible applications of genomics as in this chapter's subject: nutrigenomics. The working of the human genome and its interaction with the environment (like nutrients) are complex, which became already clear after only 22.000 genes were discovered. Interactions between genes, enormous amounts of genes reacting on one nutrient and opposing reactions of genes triggered by the same nutrient, make the challenges of nutrigenomics abundant.[2] Moreover, it is rather difficult to test the health effect of nutrients

[1] Smith, 2005

[2] Piatigorsky, 2007

and therefore certain diets and products on health because it is hard to determine the area between illness and health. Nutrients have to interact during a stage in which the body at that moment of time does not give any signals of illness yet.[3] The literature on the influence of nutrients on health is therefore replete with contradicting evidences.[4] Besides these complexities and uncertainties of the experimental results of nutrigenomics, nutrigenomics' interaction with society is largely unknown.

The issue of uncertainty of genomics is tackled earlier, e.g. from the perspective of policymaking[5], science management[6] and theory of science.[7] Ethics and politics of science have often difficulties in taking into account fundamental uncertainties because of the often irresistible urge to assess and govern innovation trajectories from an outside and not from an inside perspective. Here we will concentrate on the question how far the approach of deliberative democracy can adequately manage fundamental uncertainties of nutrigenomics, and as a consequence, more general fundamental uncertainties of science. However, we will first concentrate on nutrigenomics and in the last sections generalize our findings to science in general. Versions of the deliberative democracy approach get shape in various research and policy experiments with technological innovations, for example in GM-nation, in forms of edemocracy.[8] Although most of the studies about these trials are rather positive, the disadvantages of the deliberative approach are often discussed. As one of the disadvantages is often mentioned the long time span it takes to organize deliberations (in a broad sense) and consequently deliberative democracy tools are thought to be not really apt for handling sudden and unpredictable drawbacks and twists of innovatory trajectories. However, I will show that the opinion shaping tool of deliberative democracy gives good opportunities to handle uncertainties of innovatory trajectories, is indeed even very

---

[3] Van Roost, 2005

[4] Davis and Milner, 2004

[5] Gottweis, 2005; Pinch, 2000

[6] Pinch, 2000; Brown & Michael, 2003

[7] Brown & Michael, 2003; Jasanoff, 2005

[8] Dahlgren, 2005; Renn, 1994; Burgess, 2003; Einsiedel & Eastlick, 2000; Poortinga & Pidgeon, 2004; Understanding Risk Team, 2004

adequate for identifying and considering fundamental uncertainties and to initiate a robust tool for developing innovation trajectories.

Uncertainties galore

The short history of nutrigenomics sketched in the introduction shows us an up and down of expectations and certainties. The sociologist Merton coined the term organised scepticism[9] as specific for science, which stands for the continuing criticism that is exercised by scientists on their insights and results. However this history shows also that during the period of new and starting paradigms or disciplines, scientists have the need for an organising utopian idea that rallies their energy, directs their attention and attracts attention of funding organisations. Organised utopianism (promises) seems necessary too. The two alternate, they follow each other, and the latest phase of nutrigenomics is that organised scepticism has surpassed organised utopianism: certainty and absolute truths were destroyed or better, were unmasked as too easy (un)truths and uncertainties. The intellectual persistent dispute among competing research teams is an indication of the evolution of the scientific understanding of the world. Scientists need to have the right to make mistakes and promises; that is the main message of the scientific ethos as is laid bare by Robert Merton in his sociology of science.[10]

It turns out that (intentionally or not) false promises seem to be needed and thereby mistakes are made. The history of science is full of these mistakes, which in some cases even can become big problems or even catastrophes. Use of lead, radioactivity, DDT in agriculture and food, and the claim that all vegetable oils are healthy are examples of such serious mistakes in agriculture and food sciences.[11]

Since nutrigenomics also has to take into account non-human organisms that are involved with food intake and digesting (like bacteria), it is even more complex than genomics. These complexities affect the types of the uncertainties surrounding the different

[9] Merton, 1968

[10] Merton, 1968; Pinch, 2000

[11] Bryson, 2003

branches of nutrigenomics. The uncertainties do not rise due to discordant findings, but are more complex, as we will see.

Six types of uncertainties surrounding nutrigenomics services and products

The deliberative evaluation of the dynamics of nutrigenomics services and products has to consider at least six types of uncertainties. First, the experimental, internal, uncertainties we talked earlier about, play a role: genes can have different functions in interaction with their environment and the cause effect relations are mostly until now unknown. Second, in relation to personalized dietary advice based on personal DNA it is unknown how frequently these kinds of polymorphisms will occur. Will every individual have a certain vulnerability to a certain disease or will it only be applicable to a small group of the population. Third, uncertainties are formed by the risk analyses. Will it be possible to calculate a certain risk of getting a certain condition with a certain lifestyle? Can one exclude false positives and false negatives? How difficult will it be to separate the genetic component and the lifestyle component? How high will these risks be? Fourth, for the usefulness of such tests it is uncertain to what extent risks indications about obesity and diabetes and other vulnerabilities really influence people to live healthier lives and therefore help to prevent these conditions. How will these risks be handled by the actors? Fifth, it is uncertain how nutrigenomics products and services will be developed and used. Will it be possible to develop more effectively health-improving products? Or will nutrigenomics be used as a commercial tool? Six, one fundamental uncertainty, last but not least, is the expectations and promises made by scientists and technologists. Do these turn into 'valorisation-results' of science and technology trajectories, or do they get a totally different twist or no outcome at all?

Experimental uncertainties

The experimental, internal uncertainties of nutrigenomics are quite striking. Besides the more 'normal' uncertainties of experimenting, like the lack of standardization of data, of difficulties in establishing causal connections on the basis of sound scientific evidence, there are more fundamental uncertainties. The classical idea of one gene - one protein is now abandoned and genes encode a variety of transcripts and proteins.[12] Genes can have different functions in interaction with their environment.[13] They code for proteins that can have different functions, even opposing functions. The phenomenon of moon lighting, gene-splicing or gene-sharing allows proteins to function in distinct modes and makes the road of genotype to phenotype more complex than ever.[14] We cannot know for certain how far it ever will be known what the beneficent or maleficent effect of nutrients, foods or even diets will be. This uncertainty can be seen as a cognitive uncertainty, since we do not know what the truth is in relation to nutrients and DNA. This uncertainty could affect moral issues concerning tests that indicate lifestyle related diseases in an early stage or tests that indicate susceptibilities to lifestyle related diseases. It would also pose questions about the functional food industry and its claims. This cognitive uncertainty may therefore determine moral uncertainty. Uncertainty about the functions of genes may give rise to questions about what is the right thing to do in relation to moral questions surrounding diagnostic tests or DNA-tests and choices for others, if the usefulness of this kind of tests is uncertain.

Uncertainties of personalized dietary advice

Other cognitive uncertainties are related to this complexity of nutrient DNA interaction and linked to personalized dietary advice. There are different possibilities. One possibility is personalized advice based on a DNA test that might reveal susceptibilities to certain lifestyle related conditions so that personalized dietary advice can be given. An example of the possibility of genetic testing to predict the susceptibility to certain dietary related conditions is the current research on obesity. Jeffrey Friedman, one of the pioneers in obesity research, believes that most of the variance of obesity is due to genetic factors.[15] Recently, Meyre et al[16] have found proof that three polymorphisms for the gene that codes ENPP1 could be a marker

---

[12] Pearson, 2006
[13] Piatigorsky, 2007
[14] Sriram, 2005
[15] Friedman, 2004
[16] Meyre et al. 2005

for increased risks for obesity and diabetes. According to the BBC news this could be the first step towards a DNA-test for obesity and diabetes.[17]

Another possibility could be personalised diets based on biomarkers on DNA level. By measuring mRNA and proteins that are formed after a western diet and by measuring the dynamics of these markers, it might be possible to tell that a person is developing lifestyle related conditions in an early still reversible stage. Or it might be possible to tell that a person is vulnerable to develop these lifestyle related diseases and therefore should change his or her diet. But, such biomarkers are as yet not available, and their clinical validity and utility are still to be proven.

Uncertainties of risk analyses

Third, uncertainties are connected with risk analyses. Risk analyses are uncertain first of all because the usual translation of uncertainties in risks (with the common formula of chance times impact) is not always unambiguous, which in particular became clear in the debate about Genetic Modification.[18] Moreover, many risks with respect to health vulnerabilities do not develop into full diseases. From daily life the experience of a small illness that is healed by the body is rather common. It is interesting to notice that large-scale research in populations shows again and again that many people have had strokes or brain damages without recognizing it, and without any later problems.[19] There is quite a lot of chance that a predisease does not develop into a disease due to the normal reactions of the body. Risks are also uncertain because of the above-mentioned complexity of nutrient genome interactions.

Social and moral uncertainties

How will professional groups, like dieticians, supermarkets, general practitioners, react to the developments of nutrigenomics, its complexity and uncertainties? How will they influence its applications? Would the end-users be able to integrate nutrigenomics

---

[17] BBC News, 2005

[18] Carr, 2000

[19] Rutten, 2005; Vernooij, 2008

products and services, whatever they may be, in their daily life? Nutrigenomics is thus surrounded by internal and external uncertainties.

So a fourth type of uncertainty is formed by the reaction of people upon knowledge about uncertain risks and uncertain advice and is strongly intertwined with moral uncertainty. With respect to the usefulness of (individualized) tests it is uncertain to what extent risks indications about vulnerabilities like obesity and diabetes really influence people to live healthier lives and therefore help to prevent these conditions. Health behaviour of people is determined by a complicated set of many factors that influence the final health behaviour. Besides fear for risks, which according to most health behaviour models forms an important factor for the eventual health behaviour performed, also other factors like social aspects are significant. For example people are more inclined to agree with a statement if others in the group give this statement as well because they want to conform to the opinion of the group. Norms about how to behave are important factors in this social influence.[20] Thus this cognitive uncertainty is strongly linked to moral uncertainty in relation to what people in the future society think is the right or wrong thing to do. This also causes moral uncertainty about decision making about dietary advice or DNA tests on a personal level.

Uncertainties of products

The fifth uncertainty is connected to products. Besides the above mentioned genetic tests and biomarker tests that aim to give advice about lifestyle and nutrition German and Watzke[21] go one step further and foresee personalised foods. These authors see promises for the development of foods not only for people with health problems like lactose intolerance but also for enhancing characteristics such as musical and artistic performance. Moreover they predict a future for foods designed for different metabolism types.

How nutrigenomics products and services will be developed and used is another significant uncertainty. Will it be possible to develop

[20] Korthals, 2004

[21] German & Watzke, 2004

more effective health improving products? Or will nutrigenomics be used as a commercial tool? There are always free riders that try to get a profit with products that only in name have a health effect and there are enough consumers that are so eager to improve their health situation, that governmental regulations are necessary. Are producers going to use and are they going to be able to use nutrigenomics as a tool to produce more effective healthy products? Or will it be used as a commercial tool because of limitations in the science? Or because of moral uncertainty, since it is unclear to what extent uncertain scientific results can be used for selling products. They might be used as a commercial tool and create a hype. And they may cause a situation in which people may be held responsible or feel responsible for their health, without them being really able to take full responsibility. Also the last and sixth uncertainty that will be described here is strongly related to moral uncertainty.

Uncertainties of promises and expectations

Last but not least, the role played by scientists and technologists is important and full of uncertainties: will they act out of authenticity and integrity or will they give in to media and funding organisation pressures and use the valorisation jingle? Or, do the contexts of research funding and management allow them to be authentic and incorruptible or is it stimulated that they create a sphere of expectations?[22] Here again nutrigenomics can give examples of both types of behaviour. In the recent past, many nutritional scientists were very enthusiastic about the idea of nutrigenomics and personalised diet. The Dutch food scientist B. van Ommen started the slogan: every eater the right bite.[23] Professor W. Saris, a famous food scientist from Maastricht University said: by means of a personalized DNA chart one can predict the vulnerabilities of one person. Scientists came much earlier with these utopian ideas than mass media journalists, like Bubela and Caulfield described in other cases.[24] In 2003 The Guardian came with the article: 'First it was smart drugs. Now it's smart diets. Bruce Gierson on what your genes

[22] Borup, 2006; Cranor, 2004; Van Ommen, 2008
[23] Van Ommen, 2001
[24] Bubela & Caulfield, 2004

want you to eat'.[25] Others however, resisted the temptation to join the choir, such as Dr. Muin Khoury, director of the Office of Genomics and Disease Prevention at the US Centres for Disease Control, who stated: 'Don't Rush into Gene-Based Diets', 'Right now, no one in their right mind would offer genetic testing'.[26]

Creating unjustified expectations is a tricky issue, because it undermines the trust of the public in science and scientists. For the public trust is mostly the only type of relationship with science due to the fact that the public is not able (or better: cannot be asked to be able) to fully understand the intricacies of innovatory projects (expertise does have a function!!).

Deliberative democracy and fundamental uncertainties

How can and should deliberative democratic approaches and their participants handle the uncertainties of products and services of nutrigenomics, some of which are rather fundamental? How to select the more certain results from the hypes and the misleading ones? How to attribute responsibility to originators of messages that contain a lot of uncertainty or are even filled with hypes? How to deal with the different normative expectations of the different stakeholders? How can citizens/consumers orient themselves in this always-changing forest, similar to Macbeth's forest of Birnam Wood?

This chapter aims at spelling out the deliberative democratic meaning of the different types of uncertainties connected to nutrigenomics. We differ in our approach from the usual democratic approaches to science and technology that assume a fixed, stable and consensual position of the nutrigenomics science and after that apply their ethical principles. Here, we will explicitly start with the different types of uncertainties and complexities of nutrigenomics. Nutrigenomics is here defined as the research that uses knowledge of the genome to prevent diseases or to improve health by means of food.

Although it is unknown if the expectation will become reality that genomics will produce recommendations with respect to food based

---

[25] Guardian, 2003

[26] Bruce Grierson in *The Guardian*, May 15, 2003

on personal characteristics and if so how this knowledge will be produced, we can at least imagine relevant uncertainties with respect to these recommendations. In relation to personalized dietary advice based on personal DNA it is firstly uncertain how frequent polymorphisms (in the case of genetic testing) or metabolic disturbances (in the case of biomarker tests) will occur. Will most individuals have certain vulnerabilities to a certain disease or will personalized advice only be applicable to small groups of the population. This probably will effect moral decision making, for example with respect to discrimination and health insurance issues. Secondly, as clarified below, uncertainties connected to the complexity of DNA interaction and dietary advices are related to risk analyses.

These uncertainties pose different questions. Will it be possible to calculate a certain risk of getting a certain condition with a certain lifestyle? Or will it be difficult to separate the genetic component and the lifestyle component? How certain will this calculation of risks due to susceptibilities to nutrients be? How high will these risks be? How will these risks be handled by the actors? Uncertainty about the height of potential risks and uncertainty of the effectiveness of dietary treatments poses moral questions. Will the genetic counselor tell the worrying results of a test to the other relatives at risk? Or do other persons have to do this? Are they able to give meaning to these risks?

According to the social philosopher Zygmunt Bauman feelings of uncertainty have increased in the 20th century also because of another influence. In the modern world ethical frameworks formed by philosophy and politics were imposed on the individual by society. People therefore had fewer choices. If one was not adapted to the uniformity, one was excluded and punished. In our current post-modern society however the control by the authorities is gone because of the different nature of our society. We have more leisure time. Therefore the certainty of the uniformity and of fixed and clear-cut norms are gone. Moreover we have more freedom to choose. But we are unable to see the consequences of our choices. Norms change and because of the context of globalisation, we cannot expect that all people will have the same norms and make the same choices. On the contrary, we can only expect diversity of ethics. To overcome this feeling of uncertainty people are therefore constantly trying to create themselves. They are constantly searching for new impulses.[27]

These theories imply that our current society is led by the constant search for uncertain knowledge, frequent revelation and demolishing of facts and the constant discovery of normative disagreements. How much uncertainty with our health/body can we afford? It is not that we should get rid of nutrigenomics and refrain from applying it to obesity, but we have to cope with new grey zones and questions like: What to eat in the new situation of a long life under dominantly sedentary circumstances? How to re-establish a fit between our bodies and the world? It is not an easy thing, because it is not only food. Simultaneously the exact boundaries between food and drugs are to be rewritten, as between health and disease. There have always been unclear and sometimes fussy boundaries between the two (Hippocrates: let food be your medicine!), but because of the opportunity of prevention we now have a new case.[28]

Debates are to be organized not because it is the core of deliberative democracy to talk, but because only so all relevant viewpoints and interests vis-à-vis an issue get heard, and can be commented upon, and contribute to the meaning frames that in the end stand behind the political decisions that elected politicians have to make. Not to reach consensus and to make a decision are debates organized, but to contribute to this collective process of making up your own mind. The scenarios are desirable possibilities, not always plausible future plans, and they are made to find out what impact a certain possibility can have.[29]

On the basis of the work of Hirschman we developed in earlier work[30] the right to Voice (consultation, participation), right to Exit (the possibility to leave one technological paradigm and to enter freely another) and the right to Access (the opportunity to start something, e.g. a new technological paradigm or food style). Exit stipulates that technologies should be abandoned, at least partially, after they are used out or turn out to be disappointing, as the example of GM, nuclear energy, radioactivity, VCR or the use of chemicals in

---

27 Bauman, 1995

28 Korthals, 2008

29 Peterson, 2003

30 Korthals, 2004

agriculture learn. Voice concerns not only deliberation on technologies but also the possibility of shaping of new opinions, of looking for new vocabularies of experimenting with norms and values, and of getting the information one wants (information preferences). Access covers deliberation and experimentation: creating new boundaries, experiments with new arrangements of technologies and norms. By applying these three types of rights in concrete deliberations and consultations one can identify the opportunities for new forms of cooperation by co-managing technological alternatives and selecting ethically desirable alternatives out of the possible scenarios. We conclude that living a good life in a world with genomics products and services like high tech food and drugs, implies the deployment of procedural and substantive aspects. Both the rights of voice, access and exit and the normative values of living and cooperating together require an expanded individual consent, but not the consent of collectives. Collectives cannot vote or act, only individuals can, and all trials to let collectives act, are disguised forms of acting of certain dominant individuals. Embedding high tech foods should mean connecting them with social interactions of peaceful and meaningful actors.

With respect to genomics, suppose that indeed the first type of uncertainty is structurally the case and that indeed genes can have different (for humans positive and adverse) functions in interaction with their environment, then reflections on individual responsibilities with respect to certain vulnerabilities people might have and how they should act accordingly, are irrelevant: it is fundamentally uncertain that the supposed polymorphisms produce only vulnerabilities, they probably also produce opportunities. Moreover, the ethical question on the different health systems according to which genomics products and services can be delivered seems wrongly posed. Indeed, one could reverse the reflection: the kind of health system people choose produces a different environment and consequently different functions of genes!

Uncertainty, knowledge and social philosophy

Traditional ethics has not explicitly addressed uncertainties and how to cope with them, because it presupposes stable and fixed conditions, in which principles and norms could be upheld. However, in recent philosophy of science and social philosophy some useful ideas have been developed to cope with fundamental uncertainties without deleting them.

Certainty and uncertainty are related to either knowledge or morality or a combination of both. According to traditional philosophy of knowledge, one is only allowed (justified) to claim to know something if one also believes it and is *certain* of it. To be certain you have to have direct proof that it is.[31] Therefore to be uncertain means that one does not have all the knowledge about whether something is true. However, according to relatively recent philosophy of science (such as the theory of Karl Popper), uncertainty belongs to the core of the scientific enterprise. The community of researchers lives with uncertainty by a process of constantly reviewing, by organized criticism and by developing new paradigms. Popper states for example: 'Ethics was based upon the ideas of personal knowledge and of the possibility of reaching certainty; and therefore the idea of authority. The new ethics, by contrast is based upon the idea of objective knowledge and of uncertain knowledge' (objective knowledge is according to Popper knowledge that transgresses the individual person).[32] Cognitive uncertainty is made workable by the social and moral certainty that the scientific community eventually will eliminate false negatives and false positives.

However, according to the very influential social theory of Ulrich Beck on risk society, science is also producing more uncertainty. He claims that modern societies constantly produce new problems that originate in applying modern science, that has an analytical and reductionist character. The scientific practice in the laboratory of controlling the relevant phenomena and obtaining the relevant knowledge thoroughly depends on the extent to which one can manage to 'exclude' all the factors that are not relevant.[33] The laboratory activity of scientists includes the effort to construct closed systems *relative* to the theoretical requirement to take some variables and their interactions into account, which in turn are chosen according to the relevance in view of the problem, aim and interest of

---

31 Hospers, 1953
32 Popper, 1998, p. 63
33 Hacking, 1983

the researchers.[34] These systems are never closed as such and applying the laboratory result outside the laboratory confronts one with the excluded factors, indeed, uncertainties. When scientific results move from the laboratory to society always new problems arise that are subsequently solved with new science, that however is affected by the same reductionist character and unintended and unanticipated consequences rise again.

Uncertainties, ambiguities, ambivalent innovations

Even if the big Frankensteinian risks are not very likely, patients and consumers are still confronted with a lot of uncertainties: they experience uncertainties galore! The multiple effects of foodstuffs on the body, the unclear boundaries between health and disease, the collapse of traditional truths about healthy food - like the effects of wine, chocolate, vegetable oils, and milk - make it necessary for them to sort out the hypes from the truth and to assess the different types of uncertainties.

As a science like nutrigenomics does not even have a sieve for truth, how can the citizen/consumer know what are the hypes and what are the real uncertainties? Moreover, science in a way is constantly producing mistakes, constantly producing promises and deceptions.

Although society should allow some room for uncertainty[35] and we as individuals can afford to live with some uncertainty, the main question is how far in general and in our daily routine uncertainty can exist. The cognitive uncertainties about the connections between foodstuffs and their effects are intrinsically connected with the normative ones. Claims about the health effects of certain food stuffs, or claims about the possibility to prevent or postpone a certain disease mix up cognitive and normative claims. They also confront the patients and consumers with the question: Who to trust? Which claims are reliable and which are not? Therefore the boundaries between the dimensions of cognitive and normative uncertainties are fluid.

---

[34] Radder, 2003

[35] Douglas, 2001

Mary Douglas, in her 2001 article 'Dealing with uncertainty', asks rhetorically 'Where do we humans get our confidence in certain knowledge? The answer is cultural learning. …We create institutions that protect our valued ideas' (p. 148). But we pay a particular prize for certainty: closure of debate, taboos, correctness and the institutions that try to uphold correct behaviour. Her reflections on certainty very much resemble those of Wittgenstein in *On Certainty*. He stresses the importance of fundamental certainties that cannot be doubted, or, when doubted, get reactions like: 'this person is nuts.' He states that the human condition begins with non-doubting: 'The game of doubting itself presupposes certainty.' (section 115) A doubt requires a ground for itself, i.e. a context we trust. Doubting is only possible because we are absolutely certain of certain basic propositions, like this is my hand, or it is impossible that I was yesterday on the moon. *With respect to nutrigenomics we must decide which aspects need to be certain and which can be doubted. We must create institutions that enable us to do so: deliberations in the broadest sense.*

Co-evolution of genomics and ethics

How to organise deliberations that enable interested citizens/consumers to live with uncertainties, ambiguities, and paradoxes of genomics? How to tackle uncertainty? In accordance with the pragmatist ethics developed earlier,[36] it is ethically desirable to discuss the uncertainties, paradoxes and various scenarios in debates, that are organized on the basis of the Jamesian and Deweyian idea of inclusion: 'The course of history is nothing but the story of men's struggles from generation to generation to find the more and more inclusive order. *Invent some manner* of realizing your own ideals which also satisfy the alien demands, that and that only is the path of peace!'[37]

The American pragmatist Dewey turns the relationship between certainty and uncertainty in a different light. He concedes that the search for certainty is a dominant feeling, but he points to the negative effects of this feeling and to the lack of stimulus to learning

[36] Korthals, 2008

[37] James, 1897, p. 205

capabilities and to intellectual progress that it entails. 'As long as man was unable by means of the arts of practice to direct the course of events, it was natural for him to seek an emotional substitute; in the absence of actual certainty in the midst of a precarious and hazardous world, men cultivated all sorts of things that would give them the *feeling* of certainty.'[38] Certainties give birth to all kinds of taboos and biased judgments, but the progress of learning enables human beings to experiment and to live with uncertainties. Dewey is suggesting that as long as the context of humans is full of dangers and induces feelings of uncertainty, they are prone to look for certainty; however, if the context gives rise to feelings of certainty, individuals can feel more relaxed and give more room to feelings of uncertainty. They can even become 'happy to be uncertain'.

Ethics of uncertainty

From the point of view of both recent philosophy of science and social philosophy it is no wonder that genomics does not reduce uncertainties but even produces them. However seen from the point of view of dominant ethics, the situation is catastrophic, because the six types of uncertainties we outlined go against the grain of its most cherished ethical notions like informed consent, individual responsibility and consumer control.[39] Informed consent presupposes a constant and coherent flow of certain information that one can be informed about and then one decides. Individual responsibility presupposes clear lines of connection between agents, actions, their effects and intentions and consumer control finally requires a constant and stable personality and social context.

The second uncertainty about how frequent polymorphisms that determine certain vulnerabilities, will occur, will be used by some individuals or groups either to deny that they have these risks and for others to affirm these because they think it is to their advantage. The fifth uncertainty concerning the possibility to develop more effective health improving products or the use of nutrigenomics as purely a marketing tool, will be definitely solved in the last direction.

---

[38] Dewey, 1960, p. 33

[39] Institute for the Future, 2003

Instead of reflecting on informed consent, in these situations, constructing future scenarios involving different genomic developments (no scientifically certain diagnostics for disease prevention and no cure at all, some reliable diagnostics and cures etc.) seems more appropriate. Imaginary futures constructed by artists can be helpful as well. Last but not least, deliberations between professionals and patients/consumers/citizens are needed to find out the balance of certainty and uncertainty people can live with. Or to say the same thing differently, to sort out in a deliberative way uncertainties that are relevant and ones that are of minor importance. The American pragmatist John Dewey called those deliberations 'dramatic rehearsals'.[40]

	Issues	Tool
1.Experimental Uncertainties	Vulnerabilitiesor opportunities?	Expert deliberation Dramatic Rehearsals
2.Therapeutical Uncertainties	Where do I belong?	Deliberation Physician and consumer
3. Risks	Which risks do we really have?	General deliberation (scenarios)
4. Social	Is this fake?	Do we need the product?
5. Products	Is the claim ok?	Do we need the claim?
6. Trust	Can we trust the promises of the innovators?	Longstanding learning process

The last issue of building trust through deliberations requires special attention. Annette Baier shows in her interesting text 'Trust' (1992) that trust can be built in the absence of clear power exercise.[41] Moreover, trust is not based on a contract because of the asymmetry between the trusting and trustee partners; this counter to many who think that a new contract between science and society should be made. Indeed, even in situations where the difference between

[40] Keulartz et al. 2002; McVea, 2006
[41] Baier, 1992

experts and non-experts is not upheld anymore (e.g. in some ecological experiments), there are enough contexts in which this asymmetrical distinction still is valid. Baier discusses four principles that govern relations of trust:

M: One should not manipulate others
D: One should take due care not to allow others to form reasonable but false expectations about how one will discharge the relationship
L: One must take steps to prevent any loss that others can have by the innovation
F: One must maintain fidelity to the end (integrity).

Anyhow, in longstanding deliberative learning processes, where plays, imagination and scenario building and assessment are performed, the role of expectations and their assessment can be assessed. Not manipulating, taking care of the expectations of others, loss prevention and integrity are the four pillars of these longstanding processes. Important is also the social expansion of these deliberation processes through modern ICTechnologies (for example, through direct webbing, webrooms, chatting, virtual rooms), which is only possible when the original participants are able and willing to engage in debates with their friends and colleagues.

Conclusion

Contrary to what is often claimed, deliberative democracy approaches can indeed cope with fundamental uncertainties of genomics and present a robust way of assessing and dealing with science and technology innovation trajectories. In this paper I have shown that the six types of uncertainties connected with innovation trajectories require different types of tools of the deliberative approach but altogether they can make a substantial difference.

This is only possible due to the fact that the deliberative approach covers a wide range of tools, that dependent on the contexts can be used. Deliberative democratic tools are inclusive and verbal and non-verbal means of understanding fundamental uncertainties and make it possible to distinguish the most important ones that really hurt, from the less important ones.

Literature

Baier, A. (1992), Trust. *http://www.tannerlectures.utah.edu/lectures/documents/baier92.pdf*

Bauman, Z. (1995). *Life in Fragments: Essays in Postmodern Morality*. Oxford: Blackwell

Borup M., N. Brown, K. Konrad, H. van Lente (2006). The "sociology of expectations in science and technology", *Technology Analysis & Strategic Management*, 18 (3/4), pp.285-298

Brown, N. & M. Michael (2003). "A sociology of expectations: retrospecting prospects and prospecting retrospects", *Technology Analysis and Strategic Management*, 15, pp.3–18

Bryson, B. (2003). *A short history of nearly everything*. London: Black Swan.

Bubela, T. M. & T. A. Caulfield (2004). "Do the print media 'hype' genetic research? A comparison of newspaper stories and peer-reviewed research papers", *Canadian Medical Association Journal*," 170 (9), pp. 1399-1407

Burgess, M. M. (2003). "What difference does public consultation make to ethics?" Electronic Working Papers Series, W. Maurice Young Centre for Applied Ethics, UBC, www.ethics.ubc.ca

Carr, S. and Levidow, L. (2000). "Exploring the links between science, risk, uncertainty and ethics in crop biotechnology regulation", *Journal of Agricultural and Environmental Ethics*, 12, pp. 29-39

Cranor, C. F. (2004). "Assessing Some of the Regulatory Approaches to Transgenic Plants: What Can We Learn from the Regulation of Other Technologies?" *Environmental Biosafety Research* 3, pp.29-43.

Dahlgren, P. (2005). "The Internet, Public Spheres, and Political Communication: Dispersion and Deliberation", *Political Communication*, 22 (2), pp.147–162.

Davis, C. D., & Milner, J. (2004). "Frontiers in nutrigenomics, proteomics, metabolomics and cancer prevention". *Mutation Research*, 551, pp.51–64.

Dewey, J. (1960). *The quest for certainty*. New York: Putnam.

Douglas, M. (2001). "Dealing with uncertainty". Ethical Perspectives, 8, pp.145–155.

Einsiedel, E. F. & D. L. Eastlick (2000). "Consensus conferences as deliberative democracy – A communications perspective", *Science Communication*, 21, 4, pp.323-343

Friedman, J. F. (2004). "Modern science versus the stigma of obesity". Nature Medicine, 10, pp.563–569.

German, J. B., & Watzke, H. J. (2004). "Personalizing foods for health and delight". *Comprehensive Reviews in Food Science and Food Safety*, 3, pp.145–151.

Gottweis, H. (2005). "Governing genomics in the 21st century: Between risk and uncertainty". *New Genetics and Society*, 24(2), pp.175–194.

Guardian, The (2003, May 15). "First it was smart drugs. Now it's smart diets. Bruce Gierson on what your genes want you to eat."

Hacking, I. (1983). *Representing and intervening*. Cambridge: Cambridge University Press.

Hospers, J. (1953). *An introduction to philosophical analysis* (2nd ed.). London: Prentice-Hall.

Institute of the Future (2003). "From nutrigenomic science to personalized nutrition". http://www.iftf.org/files/deliverables/SR-93 NCNG Pers Nutri.pdf

James, W. (1897/1915). *The Will to Believe and Other Essays in Popular Philosophy,* New York: Kessinger Publishing

Jasanoff, S. (2005). "In the democracies of DNA: Ontological uncertainty and political order in three states". New Genetics and Society, 24(2), 139–156.

Keulartz, J., M. Korthals, M. Schermer & T. Swierstra (2004), "Ethics in a Technological Culture. A Programmatic Proposal for a Pragmatist Approach", *Science, Technology and Human Values*, 29, 1, pp.3-30

Korthals, M. (2004) *Before Dinner. Philosophy and Ethics of Food*, Kluwer: 2005

Korthals, M. (2008), "Ethics and Politics of Food; toward a deliberative perspective", *Journal of Social Philosophy*, vol 39, 3, pp.445-463

McVea, J. F. (2006). "Constructing good decisions in ethically charged situations: The role of dramatic rehearsal". Journal of Business Ethics, 70, pp.375–390.

Merton, (1968). "Science and the social order", in: Idem, *Social theory and social* structure, enlarged edition. New York: The Free Press.

Meyre, D., Bouatia-Naji, N., Tounian, A., Samson, C., Lecoeur, C., Vatin, V., et al. (2005). "Variants of ENPP1 are associated with childhood and adult obesity and increase the risk of glucose intolerance and type 2 diabetes". Nature Genetics, 37, pp.863–8667.

Pearson, H. (2006). "Genetics: What is a gene". *Nature*, 441, pp.398–401.

Peterson, G. D., Cumming, G. S., & Carpenter, S. R. (2003). Scenario planning: A tool for conservation in an uncertain world. *Conservation Biology*, 17, pp.358–366.

Piatigorsky, J. (2007). *Gene sharing and evolution.* Cambridge, MA: Harvard University Press. USA.

Pinch, T. (2000). The golem: Uncertainty and communicating science. *Science and Engineering Ethics*, 6, pp.511–523.

Poortinga W., and Pidgeon, N.F. (2004). "Public Perceptions of Genetically Modified Food and Crops, and the GM Nation? Public Debate on the Commercialisation of Agricultural Biotechnology in the UK." (Understanding Risk Working Paper 04-01). Norwich: Centre for Environmental Risk.

Popper, K. (1998). *The world of Parmenides: Essays on the pre-socratic enlightenment.* London: Routledge.

Putnam, H. (1995). *Pragmatism: An open question.* Oxford: Blackwell.

Radder, H. (Ed.). (2003). *The philosophy of experimentation.* Dordrecht: Springer.

Renn, O., Th. Webler & P. Wiedemann (1994), Fairness and Competence in Citizen Participation, Dordrecht: Kluwer.

Rutten, F., et al. (2005). Unrecognized heart failure in elderly patients with stable chronic obstructive pulmonary disease. European Heart Journal, 26, 1887–1894.

Smith, G. (2005). *The genomics age—how DNA technology is transforming the way we live and who we are.* New York: Amacom.

Sriram, G., Martinez, J., McCabe, E., Liao, J., & Dipple, K. (2005). "Single-gene disorders: What role could moonlighting enzymes play?" *The American Journal of Human Genetics,* 76(6), pp.911–924.

Understanding Risk Team (2004), "A Deliberative Future? An independent Evaluation of the GM Nation?" Cardiff University

Van Ommen, B., (2001), "Iedere eter de juiste hap", *Natuur en Techniek*, oktober, pp.24-28

Van Ommen, B., et al. (2008). "The challenges for molecular nutrition research 2: Quantification of the nutritional phenotype". Genes & Nutrition, 3(2), pp.51–59.

Van Roost, M. (2005). "The metabolic syndrome: A ticking time bomb". Food Engineering and Ingredients, 30(2), pp.5–7.

Vernooij, M., et al. (2008). Prevalence and risk factors of cerebral microbleeds. Neurology, 70, pp.1208–1214.

Wittgenstein, L. (1969) *On certainty*. G. E. M. Anscombe & G. H. von Wright (Eds.), (D. Paul & G. E. M. Anscombe, Trans.). Oxford: Blackwell.

13

From Prescription to Reconstruction: Opportunities for Subpolitical Choices in Biotechnological and Genomics Research

Guido Ruivenkamp & Joost Jongerden

Introduction

The relation between technology and society embodies one of the most contentious issues in contemporary culture. It was with good reason that the historian Eric Hobsbawn remarked on how '*No period in history has been more penetrated by and more dependent on the natural sciences than the twentieth century. Yet no period, since Galileo's recantation, has been less at ease with it.*'[1] Numerous studies have been conducted into the political qualities of technology over recent decades emphasizing that machines and artefacts embody specific forms of power and control. In the field of agriculture and food production, the argument was made that the development and distribution of modern agricultural technologies (such as seeds and enzymes) contributed to the transfer of power and control from the farm to other actors.[2] This process was referred to as the 'industrialization of agricultural production'.

An important characteristic of the industrialization of agricultural production was the integration of agricultural production into a production chain, which implied, a loss of autonomy at the level of the farm, with a concomitant increase in the capabilities of other actors in the chain to determine the course of the production process. This process was thus characterized by a transfer of control from the producers to the suppliers, processors and distributors, with technology playing a key role. For example, when farmers decide to use particular seeds, they also need to implement the pest and nutrient management strategies embedded in these seeds. The overall

---

[1] Hobsbawn, 1995, p. 522

[2] Hobbelink, 1987; Kloppenburg, 1988; Ruivenkamp,1989

effect of this process was a tendency for traditional management strategies to be replaced by chemical ones, making farmers increasingly dependent on the suppliers of these products for successful production and reproduction on their farms. For this reason, technology was referred to as a *politicizing product*, realizing a *control at a distance* on farming practices. [3]Therefore, it was also argued, struggles for change should not (only) be waged in the old arenas of (policy) politics, but also in these new arenas (subpolitics, Beck 1997): they needed to aim at the reconstruction of technologies and the political values they embodied. Although many studies indicated the political importance of technology, however, few were able to indicate how to make technologies more 'democratic'.

The *aim* of this contribution is to discuss the possibilities for a technology development which speaks a 'lingua democratica'. This investigation involves a critical reflection which goes between critical approaches in science and technology studies on the one hand, and experiences from civil society organizations and techno-activists attempting to reconcile technology with democratic points of reference on the other. Four aspects will be considered.

The first section describes in broad outline the *embodiment of politics within the specific development of biotechnology* as contextualized by the agro-industrial production chain. It is argued that the politics of agro-industrial biotechnology are shaped by a number of what we will refer to as *structuring trends in agricultural development* and four related *disconnection processes*. With the strengthening of these disconnection processes, not only are the local landscapes of agriculture and food production changing but also we are witnessing a transfer of agency from the agricultural producer to institutions developing technology. It will be argued that biotechnology is emptying the traditional meanings of various social categories, and that new forms of political control are emerging.[4] Therefore, we refer to biotechnology as a *politicizing vehicle reorganizing agricultural and food production.*

---

[3] Ruivenkamp, 1989; 2005

[4] Ruivenkamp, 1989; 2003

The second section discusses the opportunities for a lingua democratica within this changed context.[5] If biotechnology is a politicizing vehicle and the laboratory has become a site of making politics by different means, the scientists or the technology developers become political agents. This leads to the question whether these agents allow for a lingua democratica to be spoken within the development of biotechnology. It is argued that *contradictory developments* within the biotechnological field are, in fact, creating space for the emergence of such a lingua democratica. Four contradictory developments within actual life science technology development are broadly outlined. It is argued that these contradictory developments demand a search for workable solutions to the disconnection processes, through a social reorganization of knowledge production as well as by changing the content of the politics embodied in the biotechnological products.

The third section presents some empirical evidence which illustrates that a *multitude of voices and dialects of the oppressed are appearing*, voices and dialects expressing a quest for an alternative embodiment of politics in products and a reorganization of knowledge production. It is argued that these alternative practices – these unsilenced voices of resistance – illustrate that two types or levels of space can be distinguished which are enabling a reconstruction of technology, or at least give room for manoeuvre to this end. These two levels, moreover, operate as crucial aspects for a democratization of technology development, and give local networks opportunities to disseminate their local dialects and practices within a location specific approach of biotechnology development.

Finally, this chapter concludes by reflecting on the efforts of the social and technical activists networks trying to change the actual embodiment of politics in biotechnological products, looking at the

---

[5] A changed context in which actors still subjectively refer to their traditional social categories – as being, for example, scientists, working in public or private research institutions – while objectively they are positioned as (primarily) political agents. Elsewhere we have emphasized that the de-synchronization of the productive reorganization process from the subjective identity formation – or, the disengagement of the latter from the former – is actually an important characteristic of the changed context, and that it is precisely through this discrepancy between the subjective and objective reconstruction of identities that new forms of political control are emerging. See Ruivenkamp, 2003.

interrelation between their design of sub-political products and transformation of the social organization of knowledge development into open source approaches. The focus will be on the ways in which the diversification of biotech and genomics dialects attuned to and developed from within the perspectives of local developments can further be empowered.

The politicization of production

Scientific and technological developments do not evolve in a vacuum, but in a specific and historically shaped context of socio-economic and political circumstances. Eric Hobsbawn has described the modernization of agriculture in the 19th and 20th centuries as the single change which has most dramatically cut us forever from the world of the past – a change summarized as the *disappearance of the peasantry*,[6] and realized (increasingly) through the scale of agricultural production and extent of mechanization enabling industrial laborers (and now also service sector workers) to be supplied with (relatively) cheap food.[7]

This transformation has occurred in different regions of the world at different time, but with quite similar social consequences for landless labourers and small-scale farmers. The mechanization made agricultural labourers largely redundant and small farmers vulnerable to takeover by larger and stronger farming businesses. The widespread faith in this modernization process was so strong that agricultural policies were specifically formulated by national governments and international agencies to further this as a planned modernization.[8] The modernization did indeed bring progress and advancement to society but only in tandem with a polluting and degradation of the natural environment and destruction of the peasantry. Indeed, in regions of the world where this process was initiated first – essentially, the West – the elimination of landless laborers as a class and small-scale farming as an activity are generally regarded as fairly complete now.

---

[6] Jongerden, 2008, citing Hobsbawn 1995, p. 289

[7] Vroom, 2008; Putten, Jozien Jobse van, 1996

[8] Van der Ploeg, 2004

Van der Ploeg characterizes peasants by the way they 'actively pattern the agricultural process of production' as a self-regulated resource base and an organization of relations with markets that does not permit a commodification of all dimensions of agricultural production, and thus prevents the market from becoming hegemonic over the production process itself.[9] To sustain this peasant production system requires a permanent and dynamic articulation of social labour and agro resources. Hardt and Negri describe the peasant as characterized by his/her relationship to the land, worked upon by his/her own labour, and a direct access to the means of production.[10] A crucial characteristic of the peasant is that s/he is only partially integrated and subordinated to larger economic systems. Indeed, it is argued that, although the actual death of the peasantry is still a matter of debate, the announced death of the peasantry is inextricable linked to the idea and realization of modern society.[11] In the modern world, as it increasingly became imagined, there was no place for the peasant, whose disappearance was simply considered part of the process of modernization. This apparent death of the peasantry is linked to the birth of the farmer, a modern entrepreneur who is completely integrated into the social organization of agro-industrial production chains and whose production system is based on a maximization of commodification. The destruction of the peasantry, the argument goes, was *productive* in the sense that it was contingent on the creation of a new class of farmers who become crucial actors in and defenders of the emergence of this modern agricultural production system, deeply embedded within the industrialized and commodified economy (the peasant also interacts with markets, but with limits to the commodification of the production process). The development and design of agricultural technologies has taken place within this context, one in which 'the agricultural sector' and 'farmers and farming' are located and in which it is supposed that peasants will gradually disappear worldwide – a context which assumes and implies the social organization of the global, agro-industrial food chains.

---

[9] Van der Ploeg, 2008, p. 23

[10] Hardt and Negri, 2004, p. 116

[11] Jongerden, 2008

The development of agro-food biotechnologies under the cultural horizon of an increasing modernization of agriculture within global food chains means that the social code of these biotechnologies is strongly shaped by structural developments (trends) in the social organization of the agro-industrial production chain.[12] This social organization is described as characterized by *four disconnection processes.* First, there is the *biogenetic* disconnection of agriculture from its ecological environment, through an uncoupling of seed reproduction from its natural environment and recoupling to scientific information.[13] Next, there is the *industrial* disconnection of agriculture from food production, through the division of agricultural raw materials into food components and re-aggregation of these food components (as bases and additives like flours, fats, vitamins and sweeteners) into manufactured food products. There then follows the further *commodifying* disconnection of agricultural produce from its intrinsic food quality, through an uncoupling of the agricultural sector from the food chain and its recoupling to production chains other than food (such as energy, plastics and chemicals). Finally, there is the *ecological* disconnection of agricultural produce from human health, through an emphasis on quantity rather than quality of produce and a line of processes that destroy or deny nutritional value (which may or may not be later re-added).

The inscription of these disconnection processes – as social-technical code – in the development of agrobiotechnologies results in the emergence of biotechnological products that do not only represent but further strengthen and extend these disconnection processes. Still more seeds, enzymes, and microbiological food components are developed with a technical content (the technical code) in accordance with the social meaning biotechnology had acquired. Herbicide- and pest-resistant crops, for example, represent

[12] Ruivenkamp, 1989

[13] The genetic structure of a plant is primarily formed during a long, slow process of evolution **in** the natural environment. The genetic definition of naturally evolved vegetation is thus environmentally specific. Some plants grow well in a cool climate and others in a warm one; some plants need a lot of water, some little. Through all kinds of techniques these narrow relationships between plant evolution and the natural environment have been opened, but making the 'freed' agricultural varieties dependent instead on agricultural research generally and, in particular, the scientific information incorporated into seeds. See Ruivenkamp 1989; 2005.

and stimulate a further uncoupling of the agricultural produce from its natural (and even man-made) environment, and effectively prescribe farmers to become increasingly 'modern' (in this case, through the link to an agriculture that distances the environment, especially for the demands of the market and as directed from the laboratory). Moreover, products are developed, such as new enzymes, which enable farmers to adapt the agricultural produce to those globally organized business organization and trading patterns which reinforce the uncoupling of agricultural produce from local food consumption. The agro-food biotechnological products strengthen a dual process of uncoupling the agro- and food produce from the local context by linking it to the international market through which, subsequently it is reconnected to the local context as endpoint in the product distribution supply chain.

Biotechnological development redefines the use of the local natural and social resources to such an extent and in such a way, it has been argued, that the quantitatively increased disconnection of the agricultural produce from the local natural and social environment leads to the emergence of a qualitatively changed organization of food production.[14] This changed organization of food production, operates as a 'de-territorialization' of production, in which the local territory becomes objectified, as just another factor in the production process.[15] And the 'liberation' of production from territorial constraints leads to a loss of awareness of the relations between human settlement and the environment. Moreover the de-territorialized production creates the basis for a confidence in the construction of totally artificial environments able to remedy through technical innovations the growing crisis in the environmental and territorial systems. Within this further dual process of de- and re-territorialization, control over production is shifting from the domain of (policy) politics to that of science and technology development increasingly applied in a post-territorial context.

In short, twentieth century modernization involved a policy-facilitated industrialization of small-holder farming which disconnected agriculture from the local environment

---

[14] Ruivenkamp, 1989

[15] Magnaghi 2000; Raffestin, 1984

(deterritorialization), while recent post-modernization trends have been going in the direction of a scientifically facilitated technologization of large-scale foodstuff (and other) raw material production which re-connects agriculture to artificial environments (reterritorialization). Agriculture is being relocated and redefined. Whereas food became manufactured, now the plants and the fields themselves are being manufactured; or, the organic inputs are scientifically manipulated (plant characteristics are selected) and the sites for the first stage production process (i.e. plant growth and harvesting) are recreated (ultimately indoors, as perfectly controlled environments). And while the first move involved the exclusion of the peasant and smallholdings, the second will exclude the farmer and even the land itself. The process of commercial commodification will then be completed. It is in this development from agro science fiction to scientific reality that biotechnology and genomics research is currently positioned.

An important step is taken in the direction of this de-agriculterialization, as it may be termed,[16] when the natural environmental determination of crops is loosened. It has been emphasized that through the scientification of agricultural research,[17] increased possibilities emerge to influence the genetic structure of crops and the preservation of food products. This implies that researchers may increasingly be able to determine where, when and how different crops are cultivated and processed in different ecological environments and various geographical regions.[18] As an effect of this, the regulation of agricultural production systems will tend to shift from the policy domain to that of product development – a trend that has been characterized by referring to the products in question (seeds and enzymes) as *politicizing products*. With the scientific information that is now built into them, these products discipline cultivation and food processing practices, setting up a new system of power relationships based on the determination of agricultural production *at a distance* by these research products and in particular by their immaterial content, designed by the (immaterial) labor of researchers active within the complex international network of

[16] Taking the 'agri' (from *ager* = field) out of agriculture.
[17] Van der Ploeg, 1987
[18] Ruivenkamp, 1989

cooperative knowledge development.[19] Control of agriculture, making strategic decisions about its future direction, falls then within the domain of knowledge development. This implies that the researchers become key political actors, although they may not identify themselves as such.

This scientification process also implies that the identity of locality can no longer be exclusively defined in geographical terms, in relation to the local ways of producing food. The scientific information incorporated in the products that are used in the localities comes from 'elsewhere', and the traditional relationship of environment to plant is inverted. It is these exogenous or de-territorialized products which come to shape the identity of the locality in which they are deployed – an identity that is increasingly formed in relation to these new knowledge intensive inputs, the seeds, enzymes, biocatalysts and other biotech products in which specific (politicized) scientific information is incorporated.

Within this reorganizing production system it becomes increasingly important to consciously direct the development of the knowledge intensive, politicized products according to clearly determined goals and social objectives. Actual development of new biotech products, however, takes place within complex, international fluid networks of private/public, fundamental/applied research institutions. This implies that individual researchers loose their control over and insight into their own contributions to the immaterial (politicizing) content of the end-products. Another dual process takes place: the scientific information incorporated in specific products such as seeds and enzymes takes on an increased social relevancy, and yet, at the same time, the steering of the social dimensions becomes increasingly problematic, due - not only to the removal and distance of development (from rural farmer to urban laboratory) and other (non-location specific, profit-maximizing, etc) motivations and considerations this introduces, but also - to the sheer complexity of the organization under which the products are developed.

---

19 Ruivenkamp, 2005

Thus, we conclude, the possibility for speaking 'a lingua democratica' in the domain of agricultural biotechnology depends on:

1) The possibilities for embodying other political messages in the politicized products, on whether so-called 'subpolitical products' can be developed, e.g. with location specificity built into design rather than location being selected or adapted afterwards (according to the already designed product as a predetermined given); and on
2) Whether the social organization of knowledge development can be changed in such a way that ethical dimensions and the related questions of individual researchers can be integrated in the design of these products, e.g. with strategic decision-making about future product development taking a lower level orientation (starting from the bottom-up with local, small-holder needs, for example, rather than the top-down with national food supply requirements). This implies a quest for a manoeuvring space in which the embodied political content of the knowledge intensive products can be changed. Below, we will indicate that this space is both indicated and actually provided by contradictory developments in the life science technologies themselves.

Contradictory developments within the politicized production system

We limit ourselves to a brief sketch of four contradictory developments which are emerging within the social and technical organization of biotechnology, and which offer social technical activists opportunities to reconstruct the technology and (thereby) change the embodied political content of the biotechnological products.

First of all, there is the contradiction between the *social organization of knowledge networks* and *the private appropriation (patenting) of the products* created within these networks. The development of politicizing products is based on the collection and processing of information circulating in flexible knowledge networks. Since this information is produced collectively, it is in fact common property. This knowledge can, however, be 'privatized', generally by linking its scientific content to specific material products and in particular by denying the contribution made to it from some sources (such as indigenous knowledge systems) and privileging other, selected parts of the whole innovation chain. Major difficulties are arising in all of these three aspects. The validity of patent claims is under discussion because of insecurity about exactly which products fall under the patent rights. This has led to much litigation over the interrelation of patents. Alongside this insecurity regarding the scope of the patents, there is also insecurity in view of the pure volume of patents related to the segmented and fragmented innovation chain. For example, the Agrobacterium plant transformation method contains more than thirty patents, of which each one claims rights on certain activities within the method.[20] Finally litigation arises around the specific contribution of different groups in the development of the patented products (who has invented what).

In view of these problems, various research groups are looking for other forms of protecting technological innovations, as well as aiming at sharing data and techniques and collectively developing new biotechnological products. New forms of knowledge production are emerging as workable solutions for what are gradually becoming the unmanageable problems of the patenting paradigm. Below, we will discuss how the social reorganization of knowledge production offers new perspectives for a reconstruction of the technology.

A *second* contradiction concerns the *shift of regulatory power from the domain of politics to that of production* and the *political decay of civil society organizations.* The development, supply and implementation of politicizing products have led to a steady decline in policy-based regulation of the food-production system, and hence to a side-tracking of those social organizations that operate in the policy domain. In view of the fact that the new social organization of food production has left democratically elected institutions with very little say in this field, it is hardly surprising that (in Europe, at least) they have tended to adopt a largely hostile, defensive attitude towards the development of biotechnology.

Lobbied by agro-business and risking being seen as taking an ostrich-like position, however, governments and parliaments are under pressure to admit that biotechnology does play an important role in society, and as they come to accept the increasing marginalization of their political position due to their inability to

---

[20] Hope, 2008

influence the development of the politicizing products, these institutions do reconsider their attitude to biotechnology. This process is already under way and has led to a number of initiatives – among organizations that are not fundamentally opposed to this new form of technology – to investigate the scope for working together with critical scientists to appropriate the new technology and link it to a variety of strategies for changes targeted at specific regional developments. Critical knowledge networks have been set up, for example, to promote an alternative content for the development of biotechnology in association with researchers who have opted to work for the advancement of '*biosciences in society*'.[21] Below, we will discuss some of the ways in which these networks of social-technical activists are able to change the embodied political content in various biotechnological products

The *third* contradiction is that brought about by the *complex organization of scientific labor*, which has caused researchers to be alienated from the *social significance of their work* – at a time when this social significance is becoming extremely and increasingly important. The complex organization of the development and production set-up makes it impossible for the scientific laborers involved to maintain adequate control of and effective insight into the production process, while the politicizing products they create bring about major changes in global food production systems with concurrent effects, direct and indirect, on unknown millions of people's lives. As a result of this social contradiction built into their labor, scientists find themselves exposed to conflicting forces that can amount to a moral dilemma concerning the very nature of the work they are doing. They can resolve these issues in different ways.

Faced with this social contradiction, scientists can, of course, simply go on with their research and ignore its social implications. They may prefer not to involve themselves in the politics of their work, leaving that to others while they concentrate on the technical investigations for which they are paid. And they may, furthermore, 'buy into' what is regarded as an imperfect world but the best we

---

[21] For example, the international network, 'Tailor-Made BioTechnologies for endogenous development' (TMBT), has been set up with support from the Dutch government to contribute to the development of tailor-made biotechnology with a focus on the so-called 'developing world'.

have, assuming that they are employed, overall, for the good. Alternatively, however, they might recognize that their work is interwoven with the new bio-power system,[22] and ask themselves questions of conscience about the contribution, active or passive, they are making to the creation and maintenance of this system. They may end up questioning their personal involvement in the 'organized irresponsibility'[23] of designing politicizing products, and start to look for ways in which they can deploy their technical knowledge for the development of products of a different kind. They can try to team up with a variety of other actors, not just to regain control of the applications of their intellectual labour but also to use their work in a way directed toward serving alternative strategies of change, toward reinforcing a 'subpoliticization from below', and leading to a multiplicity of dialects of technology developments related to different work styles and lifestyles[24] within location specific contexts. Below, we will indicate some of the ways in which scientists are actively participating in changing the material and immaterial product combinations in which the technology appears.

Finally, a *fourth* contradiction within the biotechnological development opening an additional perspective for reconstructing that technology, is precisely the way in which the technology is developing. It concerns the continuous expansion of technological developments leading to *technology development as an object in itself,* which becomes exclusively *reconnected to its own technical organization.* This implies that it is no longer society that decides to use either biotechnology or genomics as instruments to realize specific social objectives, but rather the comparative lead of one of those technological fields. It may be primarily the specific supply of, say, genomics research results within a certain field that determines which 'social issues' will be tackled first, rather than any prioritization derived from social needs analysis. As an area of research proves successful, moreover, the effect of this will tend to be self-

[22] Ruivenkamp, 2005

[23] A term coined by Ulrich Beck to describe knowledge development in a 'risk society' . See Beck 1997, p. 15.

[24] See Michael Hardt and Antonio Negri (1984) for a general discussion of the multiplicity of work styles and lifestyles.

perpetuating, with tangible gains leading to more research of which some will yield further positive results opening new lines of research, and so on, until there develops a vast complex of 'self-generated' (that is, technology led) research activity. The implications of this are profound.

In the context of the advanced technological society, it is the expansion of technological means as a whole – the actual state of and potential for technological development – that decides on which social objectives will be pursued through the realization of which technologies (and when, where, how and even why). This continuously expanding front of technological progress thus becomes the key player in historical development, *reducing people's activities to the functional*, dealing with the constant adaptation of the instrumental, technological rationality (and its concomitant maintenance, administration, etc.). This implies a trend towards an 'abstract' technology development, as if 'from nowhere'.[25] This acquires an assumed scientific and mythical objectivity external to social conditions, and the technological apparatus becomes integral to the human condition. People's lives enable and further the functioning of this apparatus,[26] changing life as an end-in-itself into life as a means for the maintenance and development of the new abstractum. An increasing technologization of life takes place, which is based on using life itself as the centre of potentiality for the future development of the instrumental technological rationality. A new structure of domination is emerging – a 'biopower system' – in which the potentiality of life, the potentiality of creativity, imagination, human endeavour itself becomes the force for the further development of that abstract technological development.

However, just as the emergence of the modern farms and global food chains have not resulted in a complete banishment of the peasantry, so also has the social dominance of the instrumental rationality not led to a complete disappearance of other ends and values. On the contrary, the essential denial of humanity in this instrumental efficiency is plain to see for those who look. Millions of young people

---

25 Hubbard et al. 2004, citing Donna Harroway

26 Galimberti, 2003, p. 685

especially are 'waking up' to this, some through deep analysis but most more intuitively feeling that the abstract appearance of 'objective' science and its technological development are not separated from human beings but, on the contrary, based on the repression of the individual. The deepest of the four contradictions listed here, therefore, is now resulting in the opposite movement, 'back' towards location specific potentialities for progress with social value emphasized and inbuilt. The dominance of the modernization process and the abstract technological development 'coming from nowhere' have led, conversely, to an increase of location specific socio-biotechnological developments, including alternative product design procedures, models for organizational practices, and systems and styles of farming, which impact at the deepest levels of global human ecology. These are the practical expression of a redefined territory as (importantly) social space, one in which the oppressed but no longer silenced voices of resistance are finding their own multiple dialects for expression.

The location specific approach of life science technology development

In view of these trends towards abstract and de-localized technological developments, a variety of actors are searching for opportunities to democratize the biotechnological development. This democratic development refers to the process through which technology is brought under the control of human agency and, instead of contributing to relations of submission and to the reduction of people to objects, furthers the cause of liberation and political processes that allow people to become the subjects of their own history. In the *reconstruction of technology development*, two ways, or levels through which the creation of a maneuvering space can be distinguished. These refer to material form and human agency.[27]

[27] They are not necessarily intended to be exhaustive, or even exclusive. On the contrary we emphasize that the material form is mediated by human agency and vice versa. One might argue, for example, for alternative farming principles (organic, permaculture, etc.), as providing a reconstructive space.

The first level through which a democratization of technology development can be furthered is that of the artefacts themselves, *the material form in which technology is presented* to us and in which particular social relations are embedded, carrying specific discursive elements. The material form in which technology appears is not fixed and may be changeable; as materialized ideology, a materialization of social relations, however, it cannot be changed 'freely'.

The second level refers to human agency and the knowledge systems through which material and social transformations of nature (and living organisms) are realized.[28] At this level, the real re-design or restructuring of technologies occurs as social process – or better, it poses the challenge *to develop human agency* in knowledge systems that can facilitate the emergence of artifacts that contribute to development in which people are empowered and not subservient.

A multitude of oppressed voices and dialects are appearing in different regions of the world, expressing their desire for an alternative embodiment of politics in the products of biotechnology, and involved in a quest to change the social organization of the development of the knowledge intensive politicized products and establish new social relations in the related production of knowledge. Although these two fronts are *closely interrelated*, we will discuss them separately referring to some concrete examples from various regions.

The development of subpolitical biotech products

The reconstruction of the artifact, i.e. the combined form of im/material technology is presented here as containing two phases. First, there is a *deconstruction* of technology by exposing the social relations from within the technology emerges; and second, there is the *rewriting of the embodied political content* that the technology contains. The first part, the deconstruction, concerns the unraveling of the artifact as a materialized ideology, clarifying the historically determined social relations and trends that are shaping the

[28] Understanding 'nature', broadly, as the organic environment and its life forms that we inherit, which are not necessarily wild or pristine or somehow primordial, unaffected and unchanged by human civilization.

technology, as well as clarifying the specific kinds of social organization to which the technology contributes. The second part, the rewriting, returns to some of the issues around sharing and inclusion of stakeholders and social ends.

The act of deconstructing a technology is complicated because the technology develops under a cultural horizon,[29] which refers to the general cultural assumptions that form the unquestioned, unconsidered background to all technology.[30] Technology development takes place within a cultural context in which certain basic ideas are just taken for granted. Therefore the first step is to de-mystify or deconstruct the ideology that is incorporated in the design of the technology. We have already emphasized that the perspective of an increased modernization of agriculture within the social organization of global agro-industrial food (and non-food) supply chains represents the core belief structuring the design of biotech products, within which development is taking place. We have also argued that the greater relevance of the scientific information incorporated in the products increasingly enables scientists to determine where, when and how the agricultural produce is cultivated and processed. And we have explained how the development of the biotechnology is interwoven with four disconnection processes through which agricultural production is de-territorialized, removed, that is, from its natural environment, from local food consumption, from the food chain and from local health systems.

---

[29] Feenberg, 1999

[30] The consequence of the presence of these cultural assumptions is that it is hardly questioned whether the co-creation of technological and social developments even *can* be reshaped (as the pro/contra biotech debate manifests) let alone how such a reconstruction might take place. Indeed, one can see in the social practices of many NGOs today that technologies are rejected and/or forms of resistance *are* organized, but that the imagination to re-appropriate and modify these technologies is often lacking. Therefore, the effort to reconstruct technologies to location specific developments has to struggle on several fronts, such as in the face of those cultural accepted assumptions, (ideological front), and focusing on the unequal social relations from which the technologies emerge (techno-political front), on the specific material content (code) of the technologies (the technical front), and on the social organization in which the knowledge development takes place (the socio-political front). See Ruivenkamp, 2008a.

A re-territorialist approach is not aimed only, or even primarily, at an inversion of these systems and process, but stresses most of all *a revitalization of the local natural and social resources for development.* Therefore, this re-territorialist approach has nothing to do with conserving traditional practices or a romantic return to the past. On the contrary, it represents a dynamic, progressive way forward, towards a future of *organizing open local spaces from within endogenous developments.*[31] Philosophically, it may be characterized as a bottom-up approach, in contradistinction to the presently hegemonic top-down orientation, and one also that starts from small specifics rather than grand abstraction. It is, therefore, profoundly democratic. Concretely, re-territorialization refers to things like the efforts of various networks to restore (create and recreate) a territorial relation between the products of plant breeding and the local environment. This is done in many ways on various levels in different localities.[32]

Combining the theoretical and the practical, a general strategy may be outlined thus: instead of developing seeds (etc.) that de-connect crops from their local context and (re)connect the crop varieties to global food markets, the *location specific networks* aim at developing seeds which reconnect the seeds to the territorial potentialities for development. This is done by re-establishing the relation between local crop development and local food production and consumption. Vroom for example, has described how scientists, farmers and consumers in the Andean region are using local landraces to develop pest-resistant potato varieties, tailored to the specific food habits of the region.[33] He also refers to the efforts of scientists and farmers in India to develop a dual-purpose, early maturing sorghum variety which not only facilitates the generation of additional income for farmers but also contributes to the restoration of soil fertility.

---

[31] Van der Ploeg et al. 1995

[32] The restoration of territoriality, it should be emphasized, is not the same as the restoration of the traditional practices of any given territory. Territoriality restoration *may* involve traditional practices – these are not rejected per se, any more than are the achieved results to date of biotechnology and genomics – but neither are they privileged (it is true that traditional practices will generally be territorial – locally suitable – and thus valued of themselves, but they are not necessarily very productive, for example, or egalitarian, for that matter). See Ruivenkamp, 2008b

[33] Vroom, 2008

Another example is the development of the zinc and iron enrichment of local cowpea and maize varieties in Ghana, for which also specific supporting measures have been taken to relate the local cultivation of these crops to the local consumption of national dishes such as *waakye* and *kenkey* stimulated by a national school feeding program.

Alongside these efforts to contribute to a restoration of the territorial value of the local resources,[34] there are also efforts to restore *the agency aspect* in biotechnology development, meaning avoiding relations of submission and dependency and contributing instead to processes of empowerment. Various strategies are developed in which the industrial appropriation of biotechnological farming activities by private and public institutions is challenged. Concretely this refers to the efforts of various networks to re-appropriate or even go beyond the development of hybrid seeds, and to reorganize the global production of, for example, pesticides into territorialized productions of things like Bacillus Thuriengiensis (Bt) sprays (as has been done in the southeastern Indian state of Andhra Pradesh). Concerning the production of seeds, Kloppenburg has convincingly argued that the development of hybrid seeds has been a key aspect in the social transformation of the agricultural sector into a phase within the agro-industrial production chain. In contrast to that strategy, initiatives by national research institutions and farmers groups in Ghana focus on the development of open pollinated, enriched maize varieties which are cultivated by farmers who deliver their products at the school feeding program.[35] Also, efforts have been made to go beyond the social disadvantages of the hybrid strategy by making use of the apomictic properties of certain crops (i.e. their ability to reproduce asexually) so as to circumvent the biological protection of hybrid varieties. Hybrid seeds only give high yields during the first year of their existence. Subsequent generations no longer have this hybrid power, so farmers are forced to buy new seed every year instead of laying aside part of the harvest for sowing in the traditional way. The possibility – and desirability – of crossbreeding apomixis into e.g. hybrid maize varieties is now being

---

34 Puente, 2010

35 Ruivenkamp, 2008

investigated, as a way of gaining the benefits of the hybrids without the expensive disadvantage for farmers of regular seed purchase.[36]

Staying with the agency focus, another approach is to look not for these technical solutions as such (not directly, that is), but to aim at a social reorganization of a plant-breeding program, changing the social relations between various stakeholders. This can determine the problems that are isolated as well as how their solutions are approached (which may be technical, in the sense of genomic innovation, but maybe not). Vroom has described the emergence of an international consortium working on the development of a transgenic cabbage variety in India, in which both public and private partners reconsider their positions and renew their relations with resource-poor farmers in the region.[37] The traditional role patterns within the politicized production systems, such as the patenting of specific products or the focus on hybrid varieties, place farmers as receivers of technology (and put them under various forms of pressure to take up new products and engage in particular practices). Working under the revised re-territorializing model, these patterns are reconsidered, leading, in the Indian transgenic cabbage variety case to specific re-arrangements for the sharing of intellectual property that make the dual Bt gene construct freely accessible to national research institutions, reconsider the possibility of developing open pollinated varieties and take concrete measures to re-establish farmers as (co-) innovators and stewards for the technological developments.

The above-mentioned examples merely give a flavor of what is occurring and one direction in which re-territorialization is going. They illustrate that a multitude of oppressed dialects are giving voice to the demand for an alternative embodiment of politics in key biotechnological products, such as seeds. Another strategy for changing the political embodiment of the biotechnological product is to tackle the social organization of knowledge production.

---

[36] This type of example exhibits the type of relationship that can emerge at the present time between re-territorialization, standard biotechnology and traditional farming practice, or, how re-territorialization can provide a framework whereby the best of both the old and the modern are combined.

[37] Vroom, 2008

The social reorganization of knowledge production

The interrelation between knowledge production, protection regimes and technology development is very dynamic. Empirical evidence suggests that intellectual property rights over knowledge intensive products and processes (patents) became increasingly important during the period when techniques were emerging to modify and introduce specific genes.[38] As genomics has developed over the past two decades, however, and the focus shifted from single genes to the level of genomes – the functioning of the complete package of genetic material in an organism – practical incentives have been created *to share the inputs* involved in unveiling and interpreting the vast amounts of information produced by the products of this research. It has been argued that the enormous amount of scientific information resulting from the application of new technologies force research institutions and companies to set up partnerships for processing that information.[39] The otherwise overwhelming volume of information alone motivates for cooperation.

Research institutions are also stimulated to look for alternative protection regimes by the high transaction costs of obtaining patent protection, as indicated (above). These costs may become prohibitive due to the legal and scientific complexity of determining the scope of competing patent rights and the time required for their bundling into workable groups.[40] This issue of patenting complexity is paradigmatically exemplified by the development of the genetically engineered rice variety Golden Rice, with its 70-odd different but overlapping proprietary claims.[41]

Still more companies and research institutions follow a dual strategy in knowledge development due to the specific characteristics of the life science technologies, including, alongside the fragmented

---

[38] Graff, Rausser & Small, 2003; Ruivenkamp, 1989

[39] Deibel, 2008; Vroom 2008; Overwalle, 2007; Rai, 2007; Glorioso, 2006

[40] Deibel, 2008; Hope, 2008

[41] Nottenburg et al. 2002; Kryder et al. 2000. In this particular case, a novel way through the protection minefield was found by patenting a key technology providing access to patented ancilliary (engineering) technologies for which license had been granted for humanitarian purposes (to reach vitamin A-deficient people in developing countries).
See: http://www.goldenrice.org/Content1-Who/who4_IP.php

ownership of patented intellectual assets, the high speed of information exchange and the enormous amount of information to be processed. This dual strategy is based on sharing pools of information and techniques in the upstream pre-competitive domain alongside the downstream development of patent products.[42] This duality in types of protection is also reflected in genomics research. The size of genomics projects provides an important incentive to share the costs and energy involved, as well as to the sharing of the genomics data gained in order to interpret the vast amounts of information these projects produce. At the same time, however, the nature of those data – as a string of nucleotides – allows for the appropriation (and patenting) of interesting parts for market protection in downstream product development. Traits that are currently difficult to patent because of a lack of knowledge of responsible DNA regions or genes, will become easier to patent once the full genome sequence of the respective crop is available. In other words, companies may consider the expensive and time-consuming sequencing of entire genomes (structural genomics) to be a relatively safe activity to share with others (upstream commons), but do seem still to prefer the interpretation of the data and the development of applications (functional genomics) to be in-company.[43]

In view of these characteristics of genomics research, the question that follows regards whether the trend towards sharing data and techniques offers opportunities for developing a lingua democratica. To answer that question, it may be important to distinguish between *research commons*, which refers to resources belonging equally to or shared equally by any member of a certain community, and *open source knowledge development,* which refers to the free access to scientific information and source codes of technologies, to the practical information which enables institutions/persons to participate in and transform common knowledge development. Open source allegedly promotes innovation by making technologies available in easily modified forms, at low or no costs, and under licensing terms that allow users to make changes to the technology and to use or

---

[42] Vroom, 2008; Deibel, 2008
[43] Vroom, 2008; Deibel 2008

distribute the resulting modified versions as they choose.[44] The basic characteristic of the open-source approach of particular interest here is that diverse users are allowed to make changes in technology development which may lead to technological developments attuned to location specific needs. Therefore, it is precisely this trend towards open-source scientific approaches that we consider as a potential basis on which a lingua democratica can be formed and expressed in various local dialects of technological developments.

Conclusion

This chapter aimed to contribute to the debate of whether and how a lingua democratica can be spoken in biotechnological and genomics research. It has been emphasized that many contradictory developments within the biotechnological developments are creating a space for manoeuvre in which to try and change the embodied political content of the biotechnological products – developed within the context of the agro-industrial production chain – into a location specific orientation of agriculture and food production. This has been framed in Magnaghi's terminology as a 're-territorialization', which emphasizes the land and locality, although it is equally well named with the social emphasis of Van der Ploeg's 'repeasantization'.[45] It has also been emphasized that the specific characteristics of genomics research are stimulating a reorganization of knowledge production. Alongside a continuation and evolution of the more conventional patenting regime, this involves the emergence of a new approach to sharing data and techniques, including an alternative protection protocol, as it might be termed, which may offer opportunities to strengthen the development of open-source knowledge practices. It is

---

[44] Klein 2004. Compare this with the Golden Rice example, in which access to proprietary technologies was gained only as a one-off and under strict (yet wide-raging) conditions. The wide-ranging aspect is significant here as commercial institutions are clearly more likely to want to do good when it is relatively easy, effectively by granting the realization of (what might be seen as) a panacea, i.e. a one-size-fits-all solution.

[45] Van der Ploeg, 2008

these open-source practices that may provide a new grammar for a lingua democratica in biotechnology and genomics research.[46]

Open source knowledge practices, it is suggested, will be able to give expression to a multitude of dialects in which technologies contribute to the improvement and reuse of location specific resources for a renewed agricultural and food production system in which the disconnection processes of the food chain are reversed toward a rediscovery of the deep ecological environment (i.e. including the human ecology of local social and cultural resources). It will be crucial to investigate whether and how this changeover towards a new food production system can be empowered through a democratic biotechnological development; a democratic biotechnological development that we describe as tailored to the developmental potentialities of the location specific natural and social resources. Further research is needed to investigate the ways in which the ongoing efforts of various local networks to reconstruct agro-industrial biotechnology and genomics into tailor-made biotechnologies can be strengthened. Very clear is however already that this reconstruction is even now being realized through modifications of the embodied political content of the politicized products of biotechnology into what we have called 'subpolitical artifacts', and is beginning to be reinforced by an open source knowledge approach.

The shift towards a democratic biotechnological development may be further strengthened through the establishment of certain institutional re-arrangements. To this end, therefore, we propose at policy level to transform the actual impact-testing committees – which focus on regulating and testing biotechnological products 'after' they have already been developed[47] – into what we have called 'Open forums on technology' which reflect 'beforehand' on the politics of technological developments. These 'Open forums on technology' might be structured as a combination or synthesis of the

[46] The research group 'Critical Technology Construction (CTC) of Wageningen University has launched various PhD and post-doc research projects investigating these open source and commons practices in biotechnology and genomics (see: www.ctc.wur.nl).

[47] In the Netherlands we refer to committees such as the Committee on Genetic Modification (COGEM), and Biotechnology for Animals (CBD).

presently closed expert committees and the plethora of open public debates in which various social organizations currently participate. They may reflect on the opportunities for revisioning technology, for reopening its social code and rewriting its script. They may indicate in advance the multitude of political choices for relating new technological developments to social objectives. In this way, these Open forums on technology may become an important vehicle through which to disseminate the lingua democratica in technology development and contribute indirectly to a diversification of technology trajectories.

In other words, opportunities for a lingua democratica may be based on the combined efforts aimed at *reconstructing* the embodied political content of biotechnologies into subpolitical artefacts – as is already happened in many networks in the South – and *strengthening open source approaches* in knowledge production – as the most promising avenue toward access oriented protection protocol – and by realizing institutional re-arrangements such as the establishment of *Open forums on technology – to guide new developments in the direction of the social good rather than just respond to them as givens increasingly determined by technological development itself.* It will be through these three activities – which are no figments of imagination, but already represent concrete experiences – that a lingua democratica in technology development may be spoken, expressing a multitude of dialects and technology trajectories.

Literature

Appadurai, A. (1990). "Disjuncture and difference in the global cultural economy", in *Global Culture: Nationalism, globalization and modernity*, Featherstone, M. (ed.), London: Sage Publications.

Beck, U., Giddens, A. and Lash, S. (1994). *Reflexive Modernization. Politics, Tradition and Aesthetics in the Modern Social Order*, Cambridge: Polity Press.

Beck, U. (1997)(1992). *Shaping Technology and building society. Studies in sociotechnical change.* The MIT Press, Cambridge, Massachusetts, London, England

Chesbrough, H.W. 2003. *Open Innovation: the new imperative for creating and profiting from technology*. Boston, Massachusetts: Harvard Business School Press.

Deibel, E. (2006). "Common Genomes", in *Tailoring Biotechnologies*, Vol.2, Issue.2, pp.49-84

— (2008). "Recoding Life in Common: a critical approach to post-nature", in *Reconstructing Biotechnologies: critical social analyses*, edited by Ruivenkamp, G., Hisano, S.and Jongerden, J. Wageningen: Wageningen Academic Publishers.

Drahos, P. and Braithwaite, J. (2002). *InformationFeudalism: Who owns the knowledge economy?* London: Eartscan.

Feenberg, A. (1999). *Questioning Technology*, London, New York: Routledge.

Feenberg, A. (2005). "Critical Theory of Technology: An overview", in Ruivenkamp, G. and Jongerden, J (eds), *Tailoring Biotechnologies. Potentialities, Actualities and Spaces*, Vol.1 Issue 1,.

Gibbons, M., Limoges, C., Nowotny, H., Schwartznman, S., Scott P., and Trow. M., (1994). *The New Production of Knowledge, the dynamics of science and research in contemporary societies.* London, Thousand Oaks and New Delhi: Sage.

Glorioso, A. (2006). "Floss Methods in Biotechnology", in *First Monday.* Vol. 11, Issue 6. At: http://www.uic.edu/htbin/cgiwrap/ bin/ojs/index.php/ fm/ issue/view/200

Goodman, D., Sorj, B. and Wilkinson, J. (1987). *From Farming to Biotechnology. A theory of agro-industrial development*, Oxford: Basil Blackwell Publications.

Hardt, M. and Negri. A. (2004). *Multitude: War and Democracy in the Age of Empire*, London and New York: The Penguin Press.

Hobbelink, H. (1987). "Biotechnology: New Hope or False Promise?" *Biotechnology and Third World Agriculture.* Brussels: Intl. Coalition for Development Action.

Hobsbawn, E. (1995). *The age of extremes: The short twentieth century. 1914-1991.* New York: Penguin.

Hope, J. (2004). *Open Source Biotechnology.* PhD thesis, Research School of Social Sciences and the Faculty of Law at the Australian National University.

Hope, J. (2006). *Open source patent licenses for biotechnology: Why and how?* Paper presented at the 14th Annual Conference on International Intellectual Property Law & Policy, Fordham University School of Law, New York, April 20 & 21.

Hubbard, P., Kitchen R., and Gil, V. (2004). *Key thinkers on space and place.* London: Sage Publications.

Hughes, S. (2005). "Navigating genomes: The spaces in which genes happen", in Ruivenkamp, G. and Jongerden, J. (Eds), 2005. *Potentialities, actualities and spaces.* Tailoring Biotechnologies, Volume 1, Issue 1.

Jongerden, J. (2008). "First the peasant? Some reflections on modernity, technology and reconstruction", in Ruivenkamp, G., Hisano, S., Jongerden, J., (Eds). *Reconstructing Biotechnologies: critical social analyses.* Wageningen: Wageningen Academic Publishers.

Kloppenburg, J.R.Jr (1988), *First the Seed: The political economy of plant biotechnology, 1492-2000.* Cambridge: Cambridge University Press.

Liard, C. (2001). "Open Source in the Biosciences". At http://www. 128.ibm.com/ developerworks/linux/library/l-osbio.html (Accessed April 2006)

Louwaars, N. (2007). "Tailoring Rights Regimes in Biotechnology: introducing DRIPS next to TRIPS", in *Tailoring Biotechnologies* Vol. 3, Issue 1, pp.99-114.

Magnaghi, A. (2005). *The urban village. A charter for democracy and local self-sustainable development.* London: Zed Books.

Magnaghi, A.(2000), *Il Progetto Local,.* Torino: Bollati Boringhieri Editore s.r.l..

Magnaghi, A. (2005), "Local Self-Sustainable Development: Subjects of Transformatio", in: Ruivenkamp, G., Jongerden, J. (Eds.), *Tailoring Biotechnologies. Potentialities, Actualities and Spaces*, Vol. 1, Issue 1.

Manzini, E. (2005). "Agriculture, Food and Design: New Food Networks for a Distributed Economy", in: Ruivenkamp, G., Jongerden, J. (Eds.), *Tailoring Biotechnologies. Networks, Practices and Paradigms*, Vol.1, Issue 2.

Milberry, K. (2008). "The Wiki Way: prefiguring change, practicing democracy"', in Ruivenkamp G., Hisano S. and Jongerden J. (Eds.). *Reconstructing Biotechnologies: critical social analyses*, pp.327-344. Wageningen: Wageningen Academic Publishers

Noble, D. (1985). "Social Choice in Machine Design: the case of automatically controlled machine tools", in MacKenzie, D. and Wajcman, J. *The Social Shaping of Technology*. Philedelphia: Open University Press.

Nottenburg, C., Pardey, P.G. and Wright, B.D. (2002). "Accessing other people's technology for non-profit research", in *Australian Journal of Agricultural and Resource Economics* 46 (3), pp.389-416.

Nowotny, H., Scott P., and Gibbons, M. (2003). "Mode 2' Revisited: The New Production of Knowledge", in *Minerva* 41, pp.179–194.

O'Neill, G. (2003). "Open Source Biology's Stance Earns International Honour", in *Australian Biotechnology News*.

Puente, Daniel (2010). *Redesigning Genomics -Reconstructing Societies. Local sustainable biotechnological developments*. Zutphen: Wohrmann Print Service

Putten, Jozien Jobse, van (1996). *Eenvoudig maar voedzaam, Cultuurgeschiedenis van de dagelijkse maaltijd in Nederland*. Nijmegen. Sun Uitgeverij

Van Overwalle, G. , Van Zimmeren E., Verbeure, B., and Matthijs, G. (2007). "Dealing with Patent Fragmentation in ICT and Genetics: patent pools and clearing houses", in *First Monday*. Vol. 11, Issue 6. At: http://www.uic.edu/htbin/cgiwrap/bin/ojs/index.php/fm/article/view/1912/1794

Rai, A. (2007). Knowledge commons: The case of the biopharmaceutical industry, in *First Monday*, Vol. 11, Issue 6. At: http://www.uic.edu/htbin/cgiwrap/bin/ojs/index.php/fm/article/view/1909/1791

Rajan, Kaushik Kundar. (2003). "Genomic Capital: public cultures and market logic of corporate biotechnology", in *Science as Culture* 12, pp.87-121.

Reichman, J.H. and P.F. Uhlir. (2003). "Contractually Constructed Research Commons for Scientific Data in a Highly Protectionist Intellectual Property Environment", in *Law and Contemporary Problems*, Vol. 66.

Rinia, E.J. (2007). *Measurement and Evaluation of Interdisciplinary Research and Knowledge Transfer*. PhD thesis. Leiden: Leiden University Press

Ruivenkamp, G. (1989). *De Invoering van Biotechnologie in de Agro-Industriele Productieketen. De overgang naar een nieuwe arbeidsorganisatie* (The introduction of biotechnology into the agro-industrial chain of production. Changing towards a new labour organization). Utrecht: Jan van Arkel Uitgeverij.

Ruivenkamp, G. (2003). "Biotechnology: the production of new identities", in *Organizational relationships in the networking age. The dynamics of identity formation and bonding*. W. Koot and P. Leisink (Eds.) Cheltenham: Edward Elgar.

Ruivenkamp, G. (2003). "Genomics and food production - the social choices", in *Genes for your food - food for your genes. Societal issues and dilemmas in food genomics*. R. v. Est, L. Hanssen and O. Crapels. The Hague: Rathenau Institute.

Ruivenkamp, G. (2005). "Tailor-made biotechnologies: Between bio-power and sub-politics", in *Tailoring Biotechnologies*, Vol. 1, Issue 1.

Ruivenkamp, G. (2008). *Biotechnology in Development: experiences from the South.* Wageningen: Wageningen Academic Publishers.

Ruivenkamp, G., Hisano, S., and Jongerden, J., (Eds.). (2008). *Reconstructing Biotechnologies: critical social analyses.* Wageningen: Wageningen Academic Publishers.

Sassen, S. (2006). "Electronic networks, power and democracy", in Ruivenkamp, G. and Jongerden J. (Eds.). *Tailoring Biotechnologies,* Volume 2, Issue 2, (*Commons, bits and genes*), Wageningen-Istanbul

Schurman, R.A. and Doyle Takahashi Kelso D. (Eds.). (2003). *Engineering Trouble. Biotechnology and its discontents.* Los Angeles. University of California Press

United Nations Development Program (UNDP). (2001). *Making New Technologies Work for Human Development.* Washington: UNDP.

Van de Sande T., Ruivenkamp, G. and Malo, S. (1996). "The socio-political context", in Bunders, J., Haverkort, B. and Hiemstra, W. (Eds.), *Biotechnology: Building on Farmers' Knowledge.* London: Macmillan Education Ltd..

Van der Ploeg J.D. (1987). *De verwetenschappelijking van de landbouwbeoefening Wageningen, Landbouwuniversiteit.* Wageningen Universiteit: Departement sociale wetenschappen, Paper No. 21.

Van der Ploeg, J.D. & Van Dijk, G. (Eds.). (1995). *Beyond Modernization. The impact of endogenous rural development.* Assen: Van Gorcum.

Van der Ploeg, J.D. (2004). *De Virtuele Boer.* Assen: Van Gorcum.

Van der Ploeg, J.D. (2008). *The new peasantries, struggles for autonomy and sustainability in an era of empire and globalization.* London: Earthscan.

Virno, P. (2004), *A Grammar of the Multitude. An analysis of contemporary forms of life.* Cambridge, Massachusetts: Semiotext.

Vroom, W. (2008). *Reflexive biotechnology development. Studying plant breeding technologies and genomics for agriculture in the developing world.* Wageningen: Academic Publishers

Vroom, W., Ruivenkamp G. & Jongerden, J. (2007). "Articulating alternatives: Biotechnology and genomics development within a critical constructivist framework", in *Graduate Journal of Social Science,* Vol. 4, Issue 1. pp.11-33.

Winner, L. (1985). "Do artefacts have politics? " In MacKenzie, D., and Wajcman, J. (Eds.), *The Social Shaping of Technology.* Philadelphia: Open University Press.

14

Storytelling as a Way to Enrich Democratic Discussion. The Dutch Debate on Embryo Selection for Hereditary Breast Cancer[1]

Marli Huijer

"I didn't want to wait another few years. I felt fear and restlessness in my body. As prospective mother you have to take your responsibility."

Ellen Groenewold, BRCA-mutation carrier (25), waiting for PGD.

"Does an embryo with a gene mutation have no right to live? I once was an embryo like that. I live and I am happy."

Esther Leeninga-Tijmes, BRCA mutation-carrier (30), history of breast cancer.

"If you allow embryo selection for breast cancer, there will be many more diseases to select for. There's no end to that. That's why I am not a pro. But I understand the hardship."

Annie van der Horst-Van der Top, BRCA-mutation carrier (61), history of breast cancer, 3 adult daughters.

"What do these politicians think if they tell people to accept an embryo with a gene defect? You consciously send people with such a gene defect into death, or they become severely ill. It is discussed over the heads of the ones in question."

Jos Jansen (56), BRCA-mutation carrier, 2 affected daughters.

---

[1] First Published in BioSocieties, Edited By Nikolas Rose, Adele Clarke and Illina Singh (vol. 4, September 2009), reproduced with permission of Palgrave Macmillan.

Introduction

Embryo selection (Preimplantation Genetic Diagnosis) for hereditary breast cancer is worldwide under discussion. Debates are often polarized between advocates pleading to open the option of embryo selection and opponents demanding a ban of PGD in all cases. The former ones emphasize the principle of patient autonomy. In their view, women who are carriers of a serious hereditary disease should have the option to use PGD to protect their children. Their opponents, on their turn, fear the dangers of human enhancement. They use the slippery slope argument: Since we are all genetically at risk for diseases, in the end no one will be good enough to be born. Some opponents argue that all germinate human life deserves protection. Health should not be prioritised above the value of human life itself.

In the Netherlands, this opposition escalated in May and June 2008, shortly after the State's Secretary of Health proposed in a letter to the Cabinet to skip the ban on PGD for hereditary breast cancer. The Social-democratic party (PvdA) and the Reformatory Christian party (ChristenUnie), both participating in Dutch parliament, clashed. The ChristenUnie demanded the Secretary to immediately withdraw the letter. The content was incompatible with the most fundamental beliefs of the party. The PvdA (pro embryo selection) accused the ChristenUnie of lacking the ability to give democratic account for their arguments. The ChristenUnie reversed the argument by accusing the PvdA of undemocratic behaviour. The political controversy was mirrored in the moral, well-known, controversy between women's right to self-determination and the embryo's right to live.

Surprisingly, the matter was ultimately settled in a rather friendly atmosphere. What might have contributed to the solving of the conflict is the contribution of a third partaker, i.e. women, and some men, with a family history of hereditary breast cancer who wrote or told their personal stories in newspapers and other media. In these stories, both the principle of women's self-determination and the slippery slope argument appeared to be more ambiguous than both PvdA and ChristenUnie would initially acknowledge. For the women and men involved, self-determination was not about deciding for oneself, but about taking the responsibility for others. They were very aware of the risk of the slippery slope, and pondered explicitly about it in the talks and arguments they had with others. The main issue in their stories was not how to decide for themselves, or how to prevent the slippery slope, but how to deal responsibly with the severe human suffering they witnessed in the family or experienced personally.

In this paper, I evaluate the role of personal, real life stories in the Dutch public debate following the State Secretary's proposal to skip the ban on PGD for hereditary breast cancer. What stories and arguments did carriers of the BRCA-mutation present in the public domain? What messages did they convey? What effect did the stories have on the arguments of the other players and the following course of the discussion?

Stories of lived experience are said to evoke sympathy, to promote mutual understanding, and to engender ambiguity. Does that mean that personal stories are helpful in the solving of political conflicts? Are stories enriching democratic discussions, or should we conclude that stories are more unsettling than helpful?

Context

In the Netherlands, Preimplantation Genetic Diagnosis (PGD) – also called embryo selection - is performed since 1995. Since 2003 it is officially permitted 'if prospective parents have an elevated risk of a child with a serious genetic abnormality or disease'.[2] In 2007, this formulation turned out to be too general for the practice of clinical genetics. Clinical geneticists reported that more specific guidelines were needed, especially with respect to severe hereditary cancers. Were they allowed to select for genetic mutations that predisposed individuals to develop severe diseases in less than 100%? In addition, the Dutch Federation of Cancer Patient Associations asked the Ministry of Health to permit PGD to women with an elevated risk on a child with hereditary breast cancer.

After having consulted a group of various experts, the State Secretary of Health wrote the letter in which she proposed to skip the ban on PGD for hereditary breast cancer. Her arguments to allow PGD for hereditary breast cancer were grounded in the interests of the future child (prevent suffering) and the interests of the woman

[2] Planningsbesluit, 2003

who considers having PGD (the advantages of PGD evaluated against the risks and distress that PGD brings to her).

On 26 May 2008, the State Secretary of Health, member of the PvdA, pronounced in a television news programme that henceforth embryo selection for hereditary breast cancer would be allowed. The announcement was preceded by interviews with breast cancer patients who told about the disasters that had occurred in their life and family.[3]

In the days to follow, a storm of public protests blew up. The ChristenUnie set the debate in motion. It publicly protested against the intentional creation of embryos, of which some would be selected and others destroyed. Moreover, the allowance of PGD for breast cancer was seen as a sign to society that it is better not to live than to live with a hereditary disease. The intrinsic value of embryos and the growing tendency to enhance human life – and thus to degrade 'lesser forms' of human life - were the ChristenUnie's main objections. In the words of Henk Jochemsen, an influential advisor of the party: "Life, also of an embryo, hides a secret that calls for respect".[4] Noteworthy is that none of the politicians of the ChristenUnie explicitly mentioned religious beliefs.

In the subsequent discussion, the PvdA underscored the principle of self-determination. BRCA mutation carriers should be allowed to have PGD to protect their children (website PvdA). The alternative of terminating pregnancy after 12 weeks, which is in the Netherlands allowed if a child is carrying the BRCA mutation, was deemed a more drastic intervention than PGD.

Secular public leaders, in support of the PvdA, accused the members of the ChristenUnie of being fanatical fundamentalists who lacked the ability to give a democratic account for their arguments: they would only answer to God.[5] As a minority, it was argued, they wanted to put their will on the rest of society. Sylvain Ephimenco, a well-known newspaper columnist, wrote: "Thus, these conservatives are able to put their ideas on the rest of society and to deny women access to progress and well-being".[6] Other columnists and newspaper

---

[3] *Netwerk*, May 26, 2008

[4] Visser de, in *de Volkskrant,* June 14, 2008

[5] Etty, 2008

[6] Ephimenco, 2008

editorials as well accused the ChristenUnie of tyrannical behaviour.[7] Members of the PvdA publicly asked their party leader to leave the cabinet: "You should not feel the whip of the Christians on your naked back anymore".[8]

In response, the ChristenUnie accused the secular leaders of behaving in an intolerant, indecent and undemocratic manner.[9] "Does argumentation stop and the scolding begin if something is said out of a Christian belief?", André Rouvoet, ChristenUnie leader, wondered at his party congress.[10]

What appeared to be an insurmountable conflict was solved in about six weeks. An important drive for both the PvdA and the ChristenUnie to reach a compromise was the aspiration to uphold the governing coalition. Moreover, both parties didn't want to fuel the public indignation about politicians fighting their principles by trampling on women and couples who have to live with the burden of hereditary breast cancer.

In response to the political controversy, many women (and a few men) sent letters to newspapers or told their life stories to journalists. Between the 26th of May and the end of June, about 40 letters and interviews appeared in national and regional newspapers, representing about 17 families. Some carriers appeared more than once on stage. All interviewees presented both the hardship that their family had to face and their opinion on PGD. By portraying the tragic dilemmas they face, and by showing that they deal in very different ways with this tragedy, they challenged both the argument of self-determination and the argument of the growing tendency to enhance human life. We will first deal with BRCA mutation carriers who are pro PGD, and then with the ones who are contra.

---

[7] *NRC Handelsblad*, June 13, 2008

[8] Terstall, Duyvestein, Van Amerongen in: *Trouw*, June 13, 2008

[9] Reformatorisch Dagblad, June 5, 2008

[10] Nederlands Dagblad, June 17, 2008

Content of the stories I: Do they really know what living with cancer is?

"I can kiss that State Secretary", Caroline Haasbroek, a 36-year-old BRCA mutation carrier, exclaims in a newspaper interview two days after the State Secretary proposed to skip the ban on PGD for hereditary breast cancer. She had breast cancer, mastectomy, chemotherapy, and one ovary removed. The moment she heard that her sister had breast cancer too, she knew one thing for sure: "I love to have children, but this disease should not be transmitted. It must stop at us".[11]

Ellen Groenewold, a 25-year-old BRCA mutation carrier, is like Caroline hoping to have a child via PGD. Three years ago, her mother was diagnosed with cancer for the third time, although she had had breast amputation, chemotherapy and radiation. At that moment, Ellen wanted to have clarity about her own status. When she learnt she had the BRCA gene mutation, she opted for preventive mastectomy. In the weeks following the television interview that preceded the State Secretary's statement, several other journalists wrote down Ellen's story.[12]

Nicky Westerhof, a 26-year-old BRCA mutation carrier, tells in the same reportage that she hopes to be the first carrier who will reach the age of 50. Her grandmother died a few months before her 50th birthday, her Mum died at the age of 46. Nicky had a preventive mastectomy and would love to have children who are not affected.[13]

In the three weeks following the demand of these three women to allow PGD for hereditary breast cancer, many BRCA mutation carriers follow their example and tell their stories in national and regional newspapers. One of the claims that the storytellers make is that although the members of parliament are talking about them, they hardly have any idea what it means to be part of a family with a history of hereditary breast cancer.

The hardship that carriers, their family, relatives and friends have to face, is reason for Brigitte Wesley, diagnosed with breast cancer and

---

[11] Ligtvoet: *Parool* May 28, 2008; Noordhuis & Soest, in: *Nederlands Dagblad*, 7 June 2008; Mat, *NRC Handelsblad*, June 21, 2008

[12] *Netwerk:* May 26; Groenewold in: *Dag,* June 4; Groenwold, in *Metro,* June 10; Lucas in: *Trouw,* June 24; Mee van der, in: *Algemeen Dagblad* June 5, 2008

[13] *Netwerk,* May 26, 2008

mother of a teenager, to write in a national newspaper: 'Because of my experiences, I am surprised about the ChristenUnie's point of view … The CU places theoretical, ideological and religious beliefs immediately opposite to the cruel reality … Do they really know what living with cancer is?'.[14]

'The discussion is not about me and other carriers', writes Annemieke de Haan, a 37-year-old BRCA mutation carrier, who has been tested positive for the BRCA mutation one week earlier. 'I have three daughters of 14, 11 and 8 years. They have experienced close by what it means to have cancer. They have seen my sister's breast cancer, my nieces … I myself am a Christen. I understand people who are against PGD as a matter of principle … But I want it and I take full responsibility for my choices'.[15]

Jos Jansen, a 56-year-old BRCA mutation carrier, also calls attention to the politicians' lack of knowledge about the tragedy that families have to face. His mother, grandmother and eldest sister died of breast cancer. He has breast cancer; his youngest sister has ovary cancer. His daughters (33 and 30), who are BRCA mutation carriers, have had preventive mastectomy and ovarectomy. In the interview, they state that the way politicians discuss about the issue is inhuman.[16]

"It feels like the sword of Damocles. My life is worth living, but at the same time hard and uncertain", says Emma van Leeuwen, a 40-year-old BRCA mutation carrier. Her mother died of breast cancer. Emma was pregnant of her second daughter when she heard that her mother had the BRCA mutation. "If I had known before my pregnancies that I am a carrier, I would certainly have considered PGD".[17]

Anny van Daalen, a BRCA mutation carrier, who is also in favour of embryo selection writes: 'After I had cancer, both my breasts have been amputated. Although it is quite some time ago I still feel it to be medieval'.[18]

---

[14] Wessely in: *NRC.Next,* June 9, 2008. Republished in *Metro:* June 10, 2008

[15] Haan de, in *Trouw:* June 7, 2008; *NRC Handelsblad:* June 7, 2008; *de Volkskrant:* June 7, 2008

[16] Wissen, in *Eindhovens Dagblad:* June 14, 2008

[17] Visser de, in *de Volkskrant* June 4, 2008

[18] Daalen van, in *Trouw* June 5, 2008

'Marijke', a 62-year-old BRCA mutation carrier, whose breasts, ovaries and uterus are taken away, shares her view. 'Marijke' is mother of two daughters (36, 33) and one son (31) and has several grandchildren. She is catholic, and sure that the Lord would agree with preventing the birth of severely handicapped or diseased children.[19]

Christien, a 30-year-old BRCA carrier, whose mother died of breast cancer, whose sister had breast cancer, and she herself had breast cancer, mastectomy and chemotherapy, has the most resolute opinion about her family history. She has become pregnant via PGD in Belgium. "The rest is history", she says while proudly rubbing her bulging belly.[20]

Content of the stories II: The value of a life with cancer

Understanding the hardship of BRCA mutation carriers should, however, not be equalled with agreeing to skip the ban on PGD. In response to Ellen, Nicky and Caroline's public call to understand their wish for PGD, BRCA mutation carriers with a very different view on the matter also came to the fore.

"I follow the news on embryo selection with growing amazement", says Esther Leeninga-Tijmes, a 30-year-old mother in a national newspaper: "Does life have no value any more if you have a gene mutation that elevates the risk on cancer?" Her story tells that she developed breast cancer at an early age. She is cured after having mastectomy, chemotherapy and breast reconstruction. Her grandmother, aunt and cousin died of breast cancer. "In the beginning, the thought that my daughter might have the gene mutation was almost unbearable to me. Now I understand that twenty years from now the treatment of cancer might be much better. I was strong enough to handle breast cancer; she will be too." Due to her experience of living with breast cancer, she wonders whether women who choose for PGD are sufficiently aware of the options available to prevent and treat breast cancer.[21]

---

[19] Brand van den, in *Brabants Dagblad* June 6, 2008
[20] Schalkwijk in *AD* June 13, 2008
[21] Visser de, in *de Volkskrant* June 4, 2008

Margriet Kruyver, a 51-year-old woman who had breast cancer, shares Esther's belief that life with a gene mutation is no less valuable than a life without. 'Who dares to decide that a life like mine, which is so large and rich in experiences, should better not be lived, because of the diagnosis cancer?'[22] Her father, a gynaecologist, does not share her point of view. In a letter published in *de Volkskrant*, he writes: 'If I had not made it as an embryo, by definition I would not have cared. A more beautiful or smarter brother or sister could have been born, who might have lived a happier life than I did. Since I was already there, they did not come'.[23]

Louise van Rossum, a 56-year-old BRCA mutation carrier, mother of two adult children feels one should rather love an affected child a little extra. In a large interview in a national newspaper, she tells that she and two of her three sisters are BRCA mutation carriers. One of her sisters had breast cancer at the age of 28 and died two years later. Louise and the other sister had preventive mastectomy and ovarectomy. The option to select an embryo did not exist when Louise wanted children. In retrospective, however, she is quite sure that she and her husband would not have opted for PGD. "It wasn't easy to become pregnant, and we were in agreement that we didn't want to use all kinds of artifices. If you opt for embryo selection you cannot avoid that; an IVF procedure is required at all times." In a talk with her daughter about embryo selection, Louise's daughter brought in the argument of enhancement: "She said: I don't understand that people do that. They only want a healthy child; a child there is nothing wrong with".[24]

Annie van der Horst-Van den Top, a 61-year-old BRCA mutation carrier, mother of 3 adult daughters, has the same opinion: "If you allow embryo selection for breast cancer, there will be many more diseases to select for. There's no end to that." She had breast cancer in 1990 and 2004, her eldest sister died of breast- and ovary cancer, and her youngest sister has breast cancer. She experienced the hardship, but firmly believes that her times are in God's hand: "I can entirely depart from it".[25]

---

22 Kruyver, in *De Volkskrant* June 10, 2008
23 Kruyver, in *de Volkskrant* June 14, 2008
24 Mat, in *NRC Handelsblad* June 7, 2008
25 Noordhuis & Soest, in *Nederlands Dagblad* June 7, 2008

The same holds for Catrien van den Berg, a single 46-year-old BRCA mutation carrier, whose mother died of breast- and ovary cancer. She had preventive ovarectomy, and has regular preventive check-ups for her breast. "I am always calm if I have to be controlled at the hospital. God is my source of power." It frightens her to know that she might not have existed if embryo selection would have been possible in the past.[26]

Marijke, a 30-year-old BRCA mutation carrier, interviewed by the same journalists, is "absolutely against embryo selection", because she beliefs that we should not select human life. She had a preventive mastectomy, and has two daughters. She does not want to denounce women who choose not to have children because of what they experienced, but selection is unthinkable for her.[27]

Am Schüngel, 67-year old, knows the hardship as well. Her mother, daughter and she herself had breast cancer. Her mother survived after breast amputation and reached the age of 85. Am was 37 when she discovered cancer in her breast. It was in an advanced phase, the physicians didn't give her much hope. Yet she recovered after mastectomy. Her daughter developed the breast cancer at the age of 37. She made it thanks to breast saving surgery, chemotherapy and radiation. All three of them forego genetic testing and embryo selection. Their main argument is reluctance against the idea of malleability. In Am's words: "It is not a law that you get breast cancer. I find genetic research so calculative, it turns your life into a sad life".[28]

Pieter Klok, possible BRCA mutation carrier, hesitates whether to be tested or not, especially since he is expecting his third child. In his father's family almost all women have died around their 45th birthday. In 2006 he received a letter telling him that hereditary breast and ovary cancer runs in the family. He talks to medical experts, juggles with statistics, and decides ultimately not to talk about it anymore for at least 16 years. But then he sees a woman on television demanding the right to have her embryo tested. Should he, as a well-thinking human being and potential BRCA mutation carrier do the same as

---

26 Noordhuis & Soest, in *Nederlands Dagblad* June 7, 2008

27 Noordhuis & Soest, in *Nederlands Dagblad* June 7, 2008

28 Bles van der, in *Trouw* June 24, 2008; Schüngel-van der Haar, in *Trouw* June 12, 2008

that woman? After pondering for days, his best friend tries to project himself in Pieter's offspring. "If they would tell me: you are allowed to live, but your penis will be taken off at the age of 18, I would say: let me live. Sure. It is much too beautiful, isn't it?" It is good reason for Klok to decide not to have a test. 'The only thing I hope for is that all my children love life as much as my friend does.[29]

Understand that you do not understand

It is rather unusual that so many people take up the pen and tell their life stories in response to a political conflict. Carriers of the BRCA mutation must have felt the urge to demonstrate to politicians and the public in general the uncommonness of their situation. One of the messages carried out in the letters and interviews is that common people, politicians included, have no idea what it is to live with hereditary breast cancer.

Where the politicians' arguments pro and contra PGD depended on appeals to avoid or accept human suffering, the unique life stories of these BRCA mutation carriers demonstrated what the suffering in itself entails. The stories offer social knowledge of what the effects of the ban on PGD are on families with a history of hereditary breast cancer.

The carriers' stories attempt to provide people who do not have to live with the burden of hereditary breast cancer with understanding of the hardship and disasters they face. Simultaneously, the narratives illuminate that this understanding cannot be anything but limited. Understanding means that non-affected people understand that they cannot really understand what it is to be part of a family with hereditary breast cancer. They can't share the experiences of the carriers. The stories are written out of frustration that people seem not to understand what carrying the mutation means. "Unbelievable, that people say that carrying the mutation is not so bad," Ellen says in indignation.[30] Her story evokes sympathy, but keeps others simultaneously at a distance. By doubting whether politicians can really know who she is and bringing in her own particular story, she

[29] Klok, in *de Volkskrant* June 14, 2008

[30] Mee van der, in *Algemeen Dagblad* June 5, 2008

reveals experiences that cannot be shared by those who are not genetically at high risk but yet must understand her situation to do justice to her.

By emphasizing the exceptionality of their condition, the first group of BRCA carriers challenges the slippery slope argument, i.e. that if we allow PGD for hereditary breast cancer we will certainly take the next step and allow PGD for all kind of diseases and abnormalities. Being a carrier of the BRCA mutation is in some cases as bad as carrying the gene for Huntington's disease – in which case carriers are almost 100% sure that they will develop the disease, but not at what age. The story of 25-year-old Nicky, who hopes to be the first in the family who reaches the age of 50, is comparable to stories of carriers of the Huntington gene. In that sense, the hardship she has to face is different from the suffering that 67-year-old Am Schüngel, whose mother reached the age of 85, has experienced. Some women with a family history of breast cancer have witnessed many of their relatives dying of the disease. They feel offended by the claim that not wanting your relatives to die at an early age of breast cancer is judged as a form of enhancement of human species.

A second message that these stories convey is that each story is unique. The story of Jos Jansen, whose mother, sister, and grandmother died of breast cancer and who knows he has transmitted the gene mutation to his daughters, is entirely different from Pieter Klok's story whose children have a risk of about 12,5% to be a carrier. Continuing the ban on PGD implies that we treat all cases as equal whereas the stories make poignantly clear that no story is similar. Having breast cancer in a woman's twenties is different from having it much later. Having relatives who are cured after breast cancer is another story than having lost one's mother, sister and other female relatives due to breast cancer. Developing breast cancer before a woman starts children brings in other problems than afterwards.

A third issue the stories are plain about is that the knowledge that one of the prospective parents is a BRCA mutation carrier puts couples into a tragic situation. Whatever they choose, the outcome will not be what they want. Firstly, they may opt to forego children, thereby reducing the risk of vertically transmitting the mutation to zero. That road leads to childlessness. Secondly, they may opt for pregnancy, undergo prenatal testing and have an abortion in case of a demonstrated BRCA mutation. That road leads to the termination of a wanted pregnancy. Thirdly, they may opt to conceive a child through IVF and Preimplantation Genetic Diagnosis. The physical and psychological stress of this intervention is high; the chances of success are low. Fourthly, they can take the risk of vertical transmitting the BRCA mutation and abstain from prenatal testing or PGD. In that case, the unwanted effect is that half of the couples' children may be affected with the mutation.

As long as genetic testing for BRCA mutations did not exist, it was understandable that people just had children. Until the 1980s, it was hardly known that there could be a link between cancer and heredity.[31] Moreover, it is only since the 1960s that almost perfect contraceptive methods are available. Since the 1990s, when BRCA tests were introduced, public awareness of the genetic basis of breast cancer has increased immensely. The awareness of the hereditability of breast cancer confronts women with breast cancer in the family with a responsibility that previous generations did not experience.

The knowledge that one has a transmittable gene mutation inevitably leads to the normative question whether one is allowed to transmit the mutation to one's children. It makes people responsible for the future health of their offspring. The culturally highly valued principle of patient autonomy underscores this responsibility. "As prospective mother you have to take your responsibility," as Ellen Groenewold says. Jos Jansen also raises the question of responsibility, however more in relation to the past than to the future. He feels guilty that he is the one who has transmitted the gene defect. He knows he cannot be blamed for the situation, but still sees himself as the source of the misery. Even the potential carrier, Pieter Klok, who forgoes testing, feels that responsibility. The knowledge that breast cancer runs in the family and that he might be a carrier and thus transmit it to his children is enough reason to feel responsible for his descendants.

A last, remarkable thing that the stories show is that there is no one-to-one relationship between the life stories of BRCA carriers and their opinions about PGD. Young carriers are pro and contra, older ones are pro and contra, women with children, women without children, men, religious carriers, non-religious carriers – in all groups there are people pro and contra. The only factor that might play a

---

31 Snelders, Pieters, Meijman, 2007

role in what carriers think of having PGD is the extent to which they experienced the disaster in the family. Even for the ones at the sideline, it is understandable that Nicky wants to stop the disease whereas Am quietly accepts living with the disease.

Stories and democratic ideals

Like most old democracies, the Netherlands persistently seeks to strengthen and deepen its democratic politics. In that perspective, active participation of citizens in political deliberation is thought to be vital for the survival and further development of democracy. Citizens are encouraged to come together or meet politicians to talk about collective problems, ideals, aims and actions. But how to deal with the different speaking styles that citizens bring in? Does a theory of democracy useful to the contemporary world need a broad and plural conception of communication, including storytelling, as political theorist Iris Marion Young suggests?[32]

Young promotes a theory of 'communicative democracy' in which differences in culture, perspective and interest are taken as its starting point. She distinguishes 'communicative' from 'deliberative democracy': the former starts from differences, the latter from shared understandings and a common good. In *deliberative* democracy, discussion partners are supposed to value the common goal above their particular experiences and interests. Young argues that by asking the participants to leave behind their experiences and interests, the perspectives of the privileged are likely to dominate.[33] The opportunity to learn from the experiences and ideas of other people gets lost too.

Following Hannah Arendt's view that plurality is *the* condition of all political life,[34] Young describes *communicative* democracy as a process of mutual expression of experiences and ideas, in which citizens aim to transcend differences, thereby reaching mutual understanding. In Young's words:

---

[32] Young, 1996, pp. 132-3

[33] Young, 1996, p. 127

[34] Arendt, 1998, p. 7

> The ideal of communicative democracy...recognizes that when political dialogue aims at solving collective problems, it justly requires a plurality of perspectives, speaking styles, and ways of expressing the particularity of social situations as well as the general applicability of principles. A theory of democratic discussion useful to the contemporary world must explain the possibility of communication across wide differences of culture and social position.35

Young considers stories, but also greeting and rhetoric, to be important means of expression that help to communicate differences. Stories foster understanding across differences, they reveal particular experiences of people who live in a different social context, they evoke sympathy for the particular situation and experiences of others, they explain to outsiders why insiders value what they value, and they reveal a total social knowledge from the point of view of the insiders. Narratives can play an important role in democratic discussions, Young maintains, they help to transcend differences and conflicts, and do justice to all people involved: 'Storytelling complements arguments in an communicative democracy because it tends to be more egalitarian than typical deliberative processes'.[36]

Political theorist Ruth Lister, in her book *Citizenship. Feminist Perspectives*, uses a similar argument to include informal speaking styles, especially of women, in politics. Referring to feminist theorist Kathleen Jones, she argues that for policies to be friendly to women and the multiplicity of their interests, they have to root its democracy in experiences of women, and transform the practice and concept of citizenship to fit these varied experiences.[37]

This idea of rearticulating the relationship between formal and informal politics is challenged by political theorist Seyla Benhabib. In a critical response to Young, she states that storytelling, as an aspect of informal communication in everyday life, should not become part of our formal public reasoning. Stories are less objective and general than reasoned arguments. They bring the particularity, embodiment and subjective experiences of individuals into the public - where they

[35] Young, 1996, p. 132
[36] Young, 1996, p. 132
[37] Lister, 2003, p. 166

can be seen and heard by everybody. In her view, opening up the formal discourse on politics and law for informal modes of communication would induce arbitrariness and would limit rather than enhance social justice, because the storytellers are not asked to render an account of their story on the basis of what is commonly valued or agreed upon.[38]

Whereas Young, Lister as well as Benhabib distinguish between rational deliberation and formal modes of politics on the one hand, and storytelling and informal modes of politics on the other, political theorist Deborah Stone presents a model in which the boundaries of these two modes of politics become blurred.

In her book *Policy Paradox*, Stone argues that almost all definitions of policy problems have a narrative structure. Political arguments are 'stories with a beginning, a middle, and an end, involving some change or transformation'.[39] Politicians tell for example stories of decline, stories of helplessness and control, or blame-the-victim stories. These narratives are used to explain what the world is about, and what we have to do to solve our problems.

Stone's recognition that storytelling is part and parcel of political discourse, does not mean that all stories are equal. Some are more ambiguous than others, she writes. Ambiguity is in her view essential in politics, because it enables the transformation of individual strivings into collective aims and decisions. Ambiguity provides opportunities for change, and helps to overcome political conflicts. Unambiguous types of politics, on the contrary, are of little use in the real world: 'Without ambiguity, cooperation and compromise would be difficult, if not impossible'.[40]

Her analysis of the narrative structure of all political discourses combined with her distinction between ambiguous and unambiguous stories may help us to better understand what role stories played in the discussion on PGD.

Following Stone, it becomes clear that the ChristenUnie contributed to the discussion with a story of decline, i.e. a slippery slope story: "Things have gone wrong, and if we don't do anything, a disaster will follow." The PvdA, on its turn, favoured a story of

---

[38] Benhabib, 1996, p. 83
[39] Stone, 2002, p. 138
[40] Stone 2002, p. 157

helplessness and control: "The situation for couples with hereditary breast cancer was dramatic. We have always thought that these couple were helpless towards their fate. Now, however, let us show you that in fact we can help them". In the discussion on embryo selection, these political narratives were used to explain what PGD is about, and how we should, or should not, deal with it. Both parties strived to win the controversy by telling a clear-cut, unambiguous story.

The personal stories that the BRCA-mutation carriers introduced into the discussion on embryo selection were more ambiguous than the politicians' stories. Their stories consist of many story lines with different beginnings, different middles, and different endings. Rather than providing simple answers, they emphasize the moral complexity of their situation. A clear plot is often lacking. Moreover, their style of speaking is more emotional, and less rhetorical, it is more aimed at reaching understanding than at persuading others.[41]

Following Young's analysis of what stories *do*, and Stone's analysis of the relevance of ambiguity, we may conclude that because of their ambiguity these stories foster understanding, evoke sympathy for the particular experiences of BRCA-mutation carriers, and reveal a more comprehensive social knowledge of hereditary breast cancer from the carriers' point of view. The stories have the power to complement and open up the arguments used in the PGD discussion. Because of their ambiguity, these stories transform individual views into collective affairs, they help to transcend the controversies about PGD and to take the interests and experiences of all people involved into account.

But…

But, how can we be sure that stories evoke understanding for a specific situation? Are stories really a help in solving political conflicts? What if citizens do not understand the stories of the

---

[41] In a U.K. study on the media representations of inherited breast cancer, Henderson and Kitzinger also observed that personal stories of women with inherited breast cancer introduced soft values, uncertainty, and 'shades of grey' (Henderson & Kitzinger 1999, p. 74).

BRCA-mutation carriers? Stories have their place within the informal communication of our everyday life, but that is no reason to make them part of our public life, as Benhabib said.[42] Her argument is challenged by the stories of the BRCA-mutation carriers. If they had not brought their unique life stories into the public, politicians and citizens would never have understood what living with a BRCA mutation means. Some experiences have to be brought into public life to become known at all.

Making stories part of democratic communication or deliberation is also criticized because of the emotions aroused by the stories. These emotions would interfere with the rationality that political issues and conflicts call for. In the case of the BRCA-mutation carriers, opponents of PGD for hereditary breast cancer complained that the compassion elicited by the stories of the young female carriers who want a child via PGD undercut the rational deliberation on the issue. The opponents' complaint would have been convincing if the readers and viewers of the stories were unable to evaluate the emotions that the stories evoked. No doubt, the hardships that the carriers bring to the fore in their stories affect our emotions – but that does not prevent us from evaluating these emotions for their rationality.[43] We are affected because of the exceptionality of their hardship. This exceptionality could be an argument to allow PGD for hereditary breast cancer. But, the willingness to live with this hardship, which was also brought in by BRCA carriers, could as well be an argument to continue the ban on PGD. Moreover, emotions may not only generate new, unheard arguments or perspectives, they can also interact with arguments. The television discussion between BRCA mutation carrier Caroline Haasbroek and ChristenUnie advisor Henk Jochemsen was an example thereof.[44]

But how do we know that either powerful interest groups, like the Federation of Cancer Patient Associations or politically biased journalists, do not manipulate these stories? BRCA mutation carriers Ellen Groenewold and Caroline Haasbroek are eloquent women, who, supported by clinical geneticists and a growing number of female journalists, took the opportunity to voice their wishes in

---

[42] Benhabib, 1996, p. 83

[43] Nussbaum, 2001

[44] *Nova*: 30 May, 2008

various media. As Henk Jochemsen, advisor of the ChristenUnie complained: 'The advocates of embryo selection, especially the media, have too easily put the patients on the foreground'.[45] This assertion would wash if journalists had deliberately recruited BRCA-mutation carriers who desperately seek PGD; or, if the BRCA-mutation carriers had made a collective effort to defend their interests. Yet, the survey of the letters and interviews shows that this is not the case. Many carriers have not been interviewed by journalists, but wrote their own letters to the media. They did this mostly individually. Furthermore, several journalists have written down stories of carriers who are not in favour of PGD for hereditary breast cancer. The argument of manipulation does not hold. Due to the broad spectrum of stories and the ample participation, the storytelling of the BRCA carriers gives a representative picture of what it is to live with hereditary breast cancer.

Effect of the stories

At the end of June, the State Secretary of Health wrote a second letter on PGD, which was the outcome of the consensus reached between the parties in the cabinet (PvdA, ChristenUnie and the ChristenDemocrats). In this letter, the cabinet formulated doing justice to the complexity of individual cases as one of the main goals. Next to the values of patient autonomy and the protection of the embryo, good care is mentioned as an important value: "Good care means that caregivers focus their attention to the patient's needs and interests by surrounding them with professional care, providing understandable information and specifying the possible consequences and risks of treatment".[46]

The initial attempt to present an inventory of diseases for which PGD would be allowed, was substituted by the formulation of a set of criteria, according to which each request for PGD should be evaluated. In each case, patient and physician, assisted by a multidisciplinary team of experts, have to take into account the severity and nature of the disease, the treatment options, additional

---

45 Visser de, in *de Volkskrant* June 14, 2008

46 Bussemaker, 27 June 2008, p. 6

medical criteria and psychic and moral factors. "The careful application of these criteria enables, on the one hand, to allow PGD to couples who wish to have a child without the severe genetic disease that threatens themselves and their family. On the other hand, it helps to avoid that the area of application might shift to patients who have no high individual risk on hereditary diseases."[47]

Noteworthy in this formulation is that it does justice to the needs and interests of BRCA-mutation carriers as formulated in their stories, as to the ChristenUnie's slippery slope argument and to the PvdA's call to patient autonomy. The ChristenUnie was satisfied because the inventory of diseases for which embryo selection would be allowed was substituted with a set of meticulous criteria, which guarantees that the focus of health care will stay on treatment and cure, rather than on selection. 'The feared slippery slope is off', the party wrote.[48] The PvdA was satisfied because the options for embryo selection, and thus women's autonomy, were expanded.[49]

Although it is hard to indicate what exactly helped to reach this broader, subtle and sensitive phrasing, the stories of the BRCA-mutation carriers have undoubtedly contributed to it. In the weeks after the controversy started, both the ChristenUnie and the PvdA felt urged to respond publicly to the stories of the BRCA-mutation carriers. The ChristenUnie through Jochemsen stressed that they 'did not want to brush away the suffering of these women. It is terrible what they have to face'.[50] Party leader Rouvoet told his audience at the party conference that the ChristenUnie's favourite instrument is not the handbrake but the steering wheel: the party aims to give developments a push into the right direction.[51] State's Secretary Bussemaker (PvdA), on her turn, told afterwards that in reaching a compromise she not only had to take into account the BRCA-mutation carriers who vehemently hoped for the allowance of embryo selection, but also the carriers who were concerned about the idea that they wouldn't have been there if embryo selection had been allowed before.[52]

---

47 Ibidem, p. 9.
48 website ChristenUnie, June 27, 2008
49 *NOS journaal* June 24, 2008
50 Visser de, in *de Volkskrant* June 14, 2008
51 Nederlands Dagblad June 17, 2008
52 *Eén Vandaag* July 28, 2008

In sum, the unambiguous and principled way in which both the PvdA and the ChristenUnie started the public discussion on PGD for hereditary breast cancer came across as disrespectful after listening to the ambiguous stories of women and men who had directly experienced the hardship of living with hereditary breast cancer. Their stories helped to transform the public sphere, where politicians and the public in general act and speak, into a realm where the discussion participants were prepared to listen to each other and to reach mutual understanding.

Conclusion

Political decisions related to issues of life and death may have a profound impact on individuals' everyday life. In discussions on these issues, real life stories are able to elicit understanding and sympathy for the particularity of the other. This is especially true when political conflicts emerge about these issues. Narrative sensitivity allows politicians to apply their principles with greater sensitivity and care.

The Dutch debate on the allowance of PGD for hereditary breast cancer shows that when individuals are directly involved, they can bring their unique life stories into the public, and the fierce discussion can change into a more relaxed communication, in which arguments as well as stories and emotions have a place. Storytelling can transform the public realm into a locus where collective problems are solved by opening up the ambiguities in each other's position.

The stories of the BRCA-mutation carriers offered confidence to all discussion participants because the women (and men) involved appeared to be very aware of both the possibility of the slippery slope, and the responsibility placed on their shoulders. In some cases, the responsibility for the health of their children was reason to opt for PGD, in others the fear of slippery slope was reason to forego prenatal testing and embryo selection. The stories demonstrated that allowing PGD would not lead to a situation in which couples easily opted for embryo selection. To the contrary, more than any reasoned argument, the stories of the carriers demonstrated the moral complexity of their situation. That's a huge accomplishment in a democracy in which most political leaders and citizens prefer clear-cut positions above ambiguity.

Literature

Arendt, H. (1998) (1958). *The Human Condition.* Chicago: University of Chicago Press.

Benhabib, S. (1996). "The Democratic Moment and the Problem of Difference." In: Benhabib, S. (Ed.) *Democracy and Difference. Contesting the Boundaries of the Political.* Princeton: Princeton University Press.

Bles, Wilfried van der, "Ik heb moeite met dat maakbaarheidsidee." Interview Am Schüngel, *Trouw,* June 24, 2008.

Brand, Twan van den, "Embryoselectie? Het kan, dus grijp je kans". Interview Marijke, *Brabants Dagblad,* June 6, 2008.

Bussemaker, J., "Embryo selectie", *Netwerk televisie,* May 26, 2008.

Bussemaker, J. "Brief aan de Voorzitter van de Tweede Kamer der Staten-Generaal. Standpunt Preïmplantatie genetische diagnostiek" (PG/E-2848071), The Hague, May 2008.

Bussemaker, J., "Kabinetsstandpunt prëimplantatie genetische diagnostiek. Brief aan de Voorzitter van de Tweede Kamer der Staten-Generaal", The Hague, June 27, 2008.

Daalen, Anny van, "Erfelijke borstkanker", *Trouw* June 5, 2008.

Ephimenco, Sylvain, "Pijl en boog", *Trouw* May 31, 2008.

Etty, Elsbeth, "Krijg dan maar borstkanker". *NRC Handelsblad,* June 3, 2008, p. 7.

Groenewold, Ellen, "Dit wil je niemand aandoen", *Metro,* June 10, 2008.

Groenewold, Ellen, "Mijn kind mag al die ellende niet meemaken", *Dag,* June 4, 2008.

Groenewold, Ellen, "Dat wil je niemand aandoen", *Metro* June 10, 2008.

Haan, Annemieke de, "Sinds een week is bekend: ik ben drager van BRCA2", *Trouw,* June 7, 2008. Also published as "Ik ben drager van BRCA2 en ik was blij met de brief van Bussemaker", *NRC Handelsblad* June 7, 2008. A shorter version was published as 'Embryoselectie' *de Volkskrant* June 9, 2008.

Henderson, Lesley, Kitzinger, Jenny, (1999). "The human drama of genetics: 'hard' and 'soft' media representations of inherited breast cancer". In: P. Conrad, J. Gabe (Eds.) *Sociological Perspectives on the New Genetics.* Oxford: Blackwell Publishers, pp.59-76.

Jochemsen, Henk, *de Volkskrant,* June 14, 2008.

Klok, Pieter, "Het dna-dilemma; embryoselectie", *de Volkskrant,* June 14, 2008.

Kruyver, Margriet, "Leven met borstkanker." Brief van de dag. *de Volkskrant,* June 10 2008.

Kruyver, G. 'Diepe bewondering, Margriet', *de Volkskrant* June 14, 2008.

Ligtvoet, Floor, "Ik kan die staatssecretaris wel zoenen". Interview Caroline Haasbroek, *Parool* May 28, 2008. (Published the same day in circa 15 regional newspapers).

Lister, Ruth, (2003) *Citizenship. Feminist Perspectives.* Second Edition. New York: Palgrave/Macmillan.

Lucas, Nicole, "Mijn moeder had die keus niet". Interview Ellen Groenewold, *Trouw,* June 24, 2008.

Mat, Joke, "'Mijn zus overleed toen ze 28 was"; Het verhaal van Louise van Rossum (56) over haar borstkanker-gen', *NRC Handelsblad,* June 7, 2008.

Mat, Joke, "In de praktijk; Hoe een beslissing over embryoselectie tot stand komt", *NRC Handelsblad,* June 21, 2008.

Mee, Tonny van der. "Interview. Ellen laat haar embryo's wél onderzoeken – De zware last van een gendrager". Interview Ellen Groenewold, *Algemeen Dagblad* June 5, 2008.

Melchior, Mensje, "Ethische ruzie over acht cellen". Interview met Ellen Groenewold. *Dag* June 4, 2008.

Nederlands Dagblad, redactie politiek, "Hinderlijk Calimero Rouvoet: wij willen onze waarheid niet opleggen", *Nederlands dagblad* June 17, 2008.

Noordhuis, Petra, Soest, Aaldert van, "Borstkankergen is geen zwaard van Damocles meer". Interview Marijke. "Ik ben dubbel over embryoselectie". Interview Catrien van den Berg. "Mijn tijden zijn in Gods hand". Interview Annie van den Horst-Van den Top, *Nederlands Dagblad*, June 7, 2008.

NRC Handelsblad, "Principes in praktijk; commentaar", *NRC Handelsblad* June 13, 2008.

Nussbaum, Martha C., (2001). *Upheavals of Thought. The Intelligence of Emotions.* Cambridge: Cambridge University Press.

Planningsbesluit klinisch genetisch onderzoek en erfelijkheidsadvisering. *Staatscourant*, 23 January 2003.

Reformatorisch Dagblad, "Commentaar: Onfatsoenlijk", *Reformatorisch dagblad*, June 5, 2008.

Schalkwijk, Laura, "Gezonde baby is ook na kanker mogelijk". Interview met o.a. Christien. *AD/Algemeen Dagblad* June 13, 2008.

Schüngel-van der Haar, Am, "Dan was ik er niet geweest", *Trouw* June 12, 2008.

Snelders, S.A.M., Pieters, T., Meijman, F.J., (2003) "Medische omgang met erfelijke aspecten van kanker in Nederland, 1900-1980", *Nederlands Tijdschrift voor Geneeskunde* 151, pp.712-715.

Stone, Deborah, (2002). *Policy Paradox. The Art of Political Decision Making.* Revised Edition. New York/London: W.W.Norton & Company.

Terstall, E., Duyvestein, M., Amerongen, J. van. "PvdA moet uit het kabinet stappen." *Trouw* June 13, 2008

Visser, Ellen de, "Natuurlijk ben ik boos geweest, en verdrietig". Interview met Esther Leeninga-Tijmes, *de Volkskrant*, June 4, 2008.

Visser, Ellen de, "Vergis je niet in hoe zwaar die behandeling is". Interview met Emma van Leeuwen, *de Volkskrant*, June 4, 2008.

Visser, Ellen de, "De grenzen van God: Achtergrond Embryoselectie." Interview met Henk Jochemsen, *de Volkskrant* June 14, 2008.

Wert, G. de, Legemaate, J. *NRC Handelsblad*, June 11, 2008

Wessely, Brigitte, "Weet de CU wel wat leven met kanker is?" *NRC.Next*, June 9, 2008.

Wissen, Mariëtte van, "Laten we alsjeblieft nuchter blijven". Interview familie Jansen, *Eindhovens Dagblad*, June 14, 2008.

Young, I.M. (1996). "Communication and the Other: Beyond Deliberative Democracy." In: Benhabib, S. (ed.) *Democracy and Difference. Contesting the Boundaries of the Political.* Princeton: Princeton University Press.

Websites

http://www.**christenunie**.nl/nl/faq_embryoselectie (last version: 27-06-08).

http://www.**eenvandaag**.nl/index.php?module=PX_Story&func=view&cid=13

http://www.**netwerk**.tv/search/node/embryoselectie

http://posearch.kobala.nl/zoek?pk=nos_**NOS**nl_NEW&fz=&hm=0&omroep=NOS&zender=&thema=&sort=date&pl=25&qr=embryoselectie
http://www.**nova**tv.nl/page/zoek/embryoselectie
http://www.**pvda**.nl/renderer.do/menuId/37298/clearState/false/returnPage/37298/pageId/200021702/?q=embryoselectie

About the Authors

Alle Bruggink has had a lifetime career at DSM (chemistry, pharmaceutical intermediates, performance materials) and a part time academic position at Radboud University Nijmegen in industrial chemistry. He has been the initiator or leader of a number of national research programs in chemistry aimed at (bio)catalysis and technology for sustainability.

Paul Cilliers (‡ 2011) was professor of Complexity and Philosophy and director of the Centre for Studies in Complexity at the University of Stellenbosch, South-Africa.

Julian Cockbain (DPhil) is a Consultant European Patent Attorney and a former partner in the British patent attorney firm Dehns. He is the author of numerous academic publications on European and US patent law.

Peter Derkx is professor of Humanism and Worldviews at the University of Humanistic Studies in Utrecht, the Netherlands. His current research interests include humanist traditions, meaning in life and aging well.

Erica Haimes is Professor of Sociology and Professorial Fellow in the PEALS (Policy, Ethics and Life Sciences) Research Centre at Newcastle University, UK.

Marli Huijer is Professor of Philosophy at Erasmus University Rotterdam and Associate Professor at The Hague University of Applied Sciences (The Netherlands).

Joost Jongerden is Assistant Professor at the Sociology and Anthropology of Development sub-department of Wageningen University, the Netherlands.

Michiel Korthals is Professor of Applied Philosophy at Wageningen University and Principal Investigator at CSG Centre for Life Sciences and Society (Radboud University).

Harry Kunneman is Professor of Social Philosophy at the University of Humanistic Studies in Utrecht, the Netherlands.

Tatjana Kochetkova-Meira is lecturer of Medical Ethics at Fontys University for Applied Sciences and lecturer in philosophy at the Institute of Philosophy, Amsterdam, the Netherlands.

Anders Nordgren is Professor of Bioethics and Director of the Centre for Applied Ethics, Linköping University, Sweden.

Fernando Suárez Müller is assistant professor of social and political theory and of applied philosophy at the University of Humanistic Studies in Utrecht (The Netherlands).

Kees Pieters (DPhil) is Lector 'Mainport Information' at the Centre of Expertise Mainport Innovation of the Rotterdam University of Applied Science, which coordinates research aimed at understanding and improving issues related to the city of Rotterdam as one of the largests mainports in the world. He is also owner of Condast, dedicated to development and consultancy of data-intensive software solutions.

Guido Ruivenkamp is Associate Professor at the sub-department 'Sociology and Anthropology of Development' of Wageningen University, and Extraordinary Professor at the University of Humanistic Studies at Utrecht, the Netherlands.

Andries De Smet is a PhD researcher at the Department of Philosophy and Moral Sciences of Ghent University (Belgium). His research focuses on theories of global justice, with particular attention to versions of cosmopolitanism.

Sigrid Sterckx is a Professor of Ethics at the Department of Philosophy and Moral Sciences of Ghent University and the Vrije Universiteit Brussel, Belgium. Her current research interests include ethical, legal and political aspects of the global patent system, global justice, ethical challenges regarding genomics research and development, and global environmental ethics.

Cor van der Weele is professor of humanistic philosophy at Wageningen University. She works at the department of Communication, Philosophy and Technology.

Subject Index

www.ingramcontent.com/pod-product-compliance
Lightning Source LLC
LaVergne TN
LVHW020633300926
849161LV00004B/683